COSMIC
QUERIES

COSMIC QUERIES

StarTalk's Guide to Who We Are, How We Got Here, and Where We're Going

NEIL DEGRASSE TYSON

with JAMES TREFIL

Edited by Lindsey N. Walker

NATIONAL
GEOGRAPHIC

WASHINGTON, D.C.

Since 1888, the National Geographic Society has funded more than 13,000 research, exploration, and preservation projects around the world. National Geographic Partners distributes a portion of the funds it receives from your purchase to National Geographic Society to support programs including the conservation of animals and their habitats.

National Geographic Partners
1145 17th Street NW
Washington, DC 20036-4688 USA

Get closer to National Geographic explorers and photographers, and connect with our global community. Join us today at nationalgeographic .com/join

For rights or permissions inquiries, please contact National Geographic Books Subsidiary Rights: bookrights@natgeo.com

Interior design: Melissa Farris & Sanaa Akkach

Library of Congress Cataloging-in-Publication Data
Names: Tyson, Neil deGrasse, author. | Trefil, James, 1938- author. | Walker, Lindsey N., editor.
Title: Cosmic queries : StarTalk's guide to who we are, how we got here, and where we're going / Neil deGrasse Tyson with James Trefil ; edited by Lindsey N. Walker.
Other titles: StarTalk radio (Podcast)
Description: Washington, DC : National Geographic, [2021] | Includes bibliographical references and index. | Summary: "An exploration of some of the deepest questions about our place in the universe"-- Provided by publisher.
Identifiers: LCCN 2020038953 | ISBN 9781426221774 (hardcover)
Subjects: LCSH: Cosmology--Popular works.
Classification: LCC QB982 .T964 2021 | DDC 523.1--dc23
LC record available at https://lccn.loc.gov/2020038953

Printed in China

21/RRDH/1

*To all those who are both
curious and restless,
in search of our place
in the universe*

CONTENTS

The starry skies and bioluminescent coastline of Maine's Acadia National Park glow in this composite photograph. **Pages 2-3:** Computer simulation of two black holes colliding

AUTHOR'S NOTE

StarTalk is a multi-platform (radio, podcast, television) talk show that seamlessly blends science, comedy, and pop culture. In one version of the franchise, called "Cosmic Queries," we solicit the *StarTalk* fan base for questions on a topic, and we answer them during the show. To our surprise and delight, "Cosmic Queries" has become our most beloved format.

But there's not always room to explore the deepest questions that come our way. Like, Where did it all come from? What's it all made of? Are we alone in the universe? And, how will it all end? For that, you need a book—conceived, organized, and written using the informative but still breezy DNA of *StarTalk* itself. My co-author, academic colleague, and longtime physics educator James Trefil laid important foundations for this book, while *StarTalk*'s senior producer and head writer, Lindsey N. Walker, worked tirelessly to ensure the book reflected the editorial mission of the podcast itself.

The diversity of life on Earth: a colorized composite of scanning electron microscope views of common plant seeds

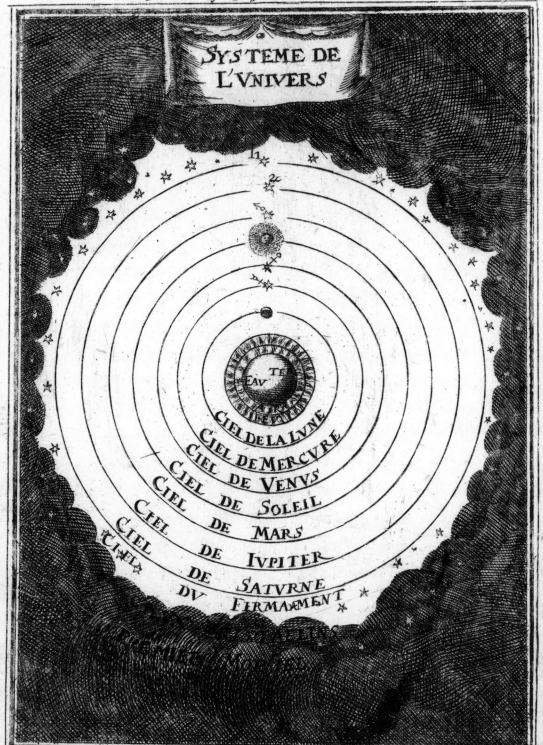

SYSTEME DE L'UNIVERS

CIEL DE LA LVNE
CIEL DE MERCVRE
CIEL DE VENVS
CIEL DE SOLEIL
CIEL DE MARS
CIEL DE IVPITER
CIEL DE SATVRNE
DV FIRMAMENT

INTRODUCTION

Hardly anyone will deny that the universe, a repository for boundless lines of inquiry, occupies a unique place in our collective core of curiosity. This fact also renders the universe a repository for our boundless collective ignorance. Small wonder that the heavens have served as the literal and spiritual home of most gods people have worshipped over the millennia. A common task of deities is to control all that seems mysterious to our mortal minds and out of control to our mortal bodies.

In the gulf between the depths of human curiosity and the limits of human ignorance resides a series of questions, some of which we all have asked and all of which some of us have asked. Not all have answers. For those that do, our answers may be incomplete or inadequate. For the remaining queries, we can look around on Earth and up into the heavens to declare with confidence, and a bit of pride, that at least some of the universe is knowable to the human mind. But we must also humbly recognize that as the area of our knowledge grows, so too does the perimeter of our ignorance.

Cosmic Queries will feed your curiosity with the deepest questions anybody has ever asked about our place in the universe. But these pages will also dip you into the eddies of our uncertainties and dangle you by your ankles above the gaps of our knowledge. Why? Because therein lies the true source of curiosity and wonder: the not knowing—coupled with its only antidote, the need to know, empowered by the methods and tools of science applied to the cosmic frontier.

A diagram created for a 1719 book shows the ancient Earth-centric worldview.

WHAT IS OUR UNIVERSE?

PLACE IN THE

1

Isaac Newton and Aristotle walk into a bar. They're engaged in an ongoing debate over what is actually going on when an object falls to Earth. Both imagine the scene, but they see completely different things.

In Aristotle's world, everything is made from the four basic elements of earth, air, fire, and water. The object, made of earth and not any of the other three elements, has an innate desire to seek the center of the universe, which—in Aristotle's view—is the same as the center of Earth. It was, after all, self-evident to him that all heavenly bodies orbited Earth, which was itself stationary. The object, then, was compelled by its inner nature to fall.

Newton doesn't care what the object is made of, only that it has mass. He knows that Earth exerts a gravitational force on every object at its surface. His law of universal gravitation tells him that anything dropped to Earth's surface will fall because of this force.

He also knows that the same force, extending out into space,

For millennia, we have sought to understand our place
in the cosmos.

keeps the Moon in its orbit, and that without the constant tug of gravity the Moon would fly off into space.

Aristotle orders a retsina. Newton orders a stiff mead. Over their drinks, they debate which view is right. Newton proposes a simple test: In his theory, neglecting air resistance, all objects dropped to Earth's surface will fall at the same rate. For Aristotle, a bigger object has more "earth element" than a smaller one, and therefore will fall faster, in proportion to how much earth element it contains. They ask the bartender for a penny and an expensive bottle of bourbon, and discover that both, although of very different mass, do indeed fall at the same rate. Newton points out that testing our ideas against nature is the core of the scientific method, a technique that has led to profound changes in the human condition through the search for objective truths and an understanding of our place in the universe.

Aristotle pays for the drinks and the broken bottle of bourbon.

IS EARTH A PLANET?

The cosmology developed by the ancient Greeks dominated thinking about our place in the universe for more than a thousand years. It taught that Earth was the unmoving center of the cosmos, the home of all life, and that all heavenly bodies, including the Sun and the stars, moved around us. Furthermore, it assumed that any imperfections on Earth do not extend into the heavens. The Sun and Moon were considered unblemished globes—the crystalline structures that carry the planets' perfect spheres rolling within other perfect spheres like invisible nesting dolls. The heavens were seen as different from Earth, made of different stuff and operating according to different laws. Earth wasn't really part of the cosmos until Isaac Newton healed the rift between them, and our own planet came to be seen as part of the universe.

THE MOST FAMOUS FAILED EXPERIMENT

Aristotle and Isaac Newton did have something in common: They believed that ether—a mysterious, invisible substance—permeated all empty space. Until the late 19th century, physicists (sensibly) presumed that, because vibrating sound waves require a medium such as air through which to propagate, light must require a medium as well—a luminiferous ether. Centuries of great thinkers had clung to the ether to help explain the inexplicable. Aristotle claimed the celestial bodies orbit in transparent crystalline spheres, with ether filling the gaps between. Isaac Newton proposed that a continual flow of ether toward Earth was the mechanism of gravity. The French mathematician René Descartes posited that invisible forces like magnetism and the tides tug and push on the ether.

But in 1887, the chemist Edward Morley and the physicist Albert Michelson provided the first compelling evidence against the idea. They reasoned that if ether permeated the space around us, then Earth's motion through it should be detectable via changes in the speed of light when measured as it moved in the same direction as Earth, compared with its speed as it moved in a direction opposite to Earth's. By analogy, if you watch a train go by and you measure the speed of a ball tossed forward from the train, what you actually get is the speed of the ball plus the speed of the train. And if the ball is thrown backward, you get the speed of the ball minus the speed of the train. Would light behave the same way?

To answer this question and analyze light beams, Michelson invented the interferometer, the most precise instrument of the time. No ether showed up. The speed of light was the same no matter which way it moved, relative to Earth. This "failed" experiment transformed science entirely and ultimately led to the discovery of special relativity.

An interferometer as designed by Michelson

In A.D. 150 Claudius Ptolemy, an Alexandrian philosopher-mathematician, articulated the ultimate Greek view of the universe. Like much of Greek science, his ideas took a roundabout journey into the curriculum of medieval universities in Europe, first translated into Arabic at the House of Wisdom in Baghdad and then later brought to Spain by crusaders and translated into Latin, the language of academic scholarship. The significance and influence of Ptolemy's book was revealed by the title the Arabs gave it, which persists to this day: *Almagest*, meaning "the greatest."

The Celestial Atlas, or The Harmony of the Universe, published in Amsterdam in the mid-1600s, portrayed the Ptolemaic universe, a system of concentric planets orbiting Earth.

A lot was going on in Ptolemy's geocentric model. The planets moved within the crystalline spheres, modified by subcycles, modified by sub-subcycles, collectively called epicycles. By adjusting the revolution speeds of all these spheres and epicycles, Ptolemy accounted for centuries of observations by Greek and Babylonian astronomers who preceded him. He could also predict eclipses and other celestial events. The system flat out worked; no wonder 1,500 years would pass before his Earth-centered model was seriously challenged.

In the Greek scheme, there were seven wanderers in the heavens: Mercury, Venus, Mars, Jupiter, Saturn, the Sun, and the Moon. The Greek word for "wanderer" is *planetes*. Because Earth was not visible in the sky, Earth was not deemed a wanderer, or planet. In this scheme, Earth is not carried along on crystal spheres—in fact, it's not carried along at all.

To the ancient Greeks, Earth is the unmoving center of the cosmos, the home of all life. (Remember that to Aristotle, the falling bourbon bottle was seeking this unmoving center.) What we call extraterrestrial life today had no place in their world. Any existence beyond Earth, anything we today would call an exoplanet, required another "Earth" surrounded by its own crystal spheres—another entire cosmos. If such an entity existed, they argued, how could an object like the falling bottle make a decision as to which center to seek out? Clearly, they argued, there can only be one center, one Earth, one cosmos.

Earthrise, December 24, 1968—a historic photograph taken during Apollo 8, the first crewed mission to the moon. Today's observations from space confirm the science of centuries past.

ASTRONOMY WITH A STICK

We don't know what you learned in third grade, but let's get one thing straight. Nobody with any education in the 15th century believed Earth was flat or that Christopher Columbus would fall off the edge if he sailed too far.

Ptolemy devoted a section in his *Almagest* to the proposition "That Also the Earth, Taken as a Whole, Is Sensibly Spherical." Among other things, he noted that solar eclipses occurred at different times of day in different places around the Mediterranean, whereas if Earth were flat, they would occur at the same time everywhere. Also, Earth's shadow on the Moon during lunar eclipses is always circular. Turns out, the only shape in the universe that casts a circular shadow, no matter the angle of sunlight, is a sphere. Ptolemy further documented how a ship

ADDING EPICYCLES

What really happens when Mercury is in retrograde? Contrary to what astrologers might tell you, nothing, because Mercury isn't actually moving backward in space. It just looks that way due to Mercury's motion relative to Earth around the Sun, like boarding a train and seeing an adjacent train move backward—only to realize that it is you who are moving forward.

But in Ptolemy's time, this apparent backward motion of planets required an active explanation that made sense within the Earth-centered model of the universe. To accommodate this periodic backward motion of planets, astronomers added smaller spheres, called epicycles, into their system. Ultimately, the Sun-centered model of the universe simplified the system, naturally accounting for retrograde motion and many other phenomena observed in the sky.

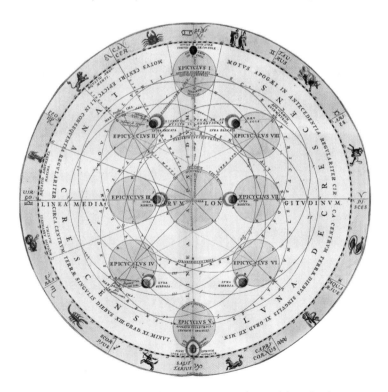

Epicycles within epicycles: Ever more complex models evolved.

> **Neil deGrasse Tyson** ✔
> @neiltyson
>
> It continues to be true that Flat-Earthers have supporters all around the globe.
>
> 💬 2.5K ↻ 12K ♡ 92.3K 4:27 PM · Oct 30, 2019

sailing away from shore would go "hull down"—that is, the hull disappears first below the horizon while the masts are still visible: evidence that the ship is sailing over the curve of Earth.

To these demonstrations, today we can add the Rose Bowl proof, which goes like this: People on the east coast of the United States who watch the Rose Bowl game being played in California see a stadium bathed in late afternoon sunlight, while outside their own windows, it's dark. If Earth were flat, darkness would fall everywhere at the same time. So both the ancients and modern football fans have ample evidence that Earth is a sphere.

Several centuries before Ptolemy, the philosopher Eratosthenes of Syene not only accepted that Earth was a sphere, but actually came up with a pretty good way of measuring its circumference. This was over a millennium ago, before the invention of the telescope and before much of anything existed in the way of astronomical instruments. His work was a prime example of "doing astronomy with a stick."

Eratosthenes knew that at noon on June 21 (summer solstice), the Sun was directly overhead, allowing its light to reach all the way to the bottom of a deep well in his hometown of Syene, near present-day Aswan, Egypt. At the same time, he measured the length of a shadow cast by a pillar, farther north in Alexandria.

To uneducated ancient observers, the night sky was like the interior of a planetarium dome, a surface speckled with brilliant objects. Stars and planets were on the sky, not in the sky.

A ROYAL JUDGMENT

Alfonso the Wise, king of Castile, upon being introduced to Ptolemy's model of the solar system, allegedly remarked: "Had I been present at the Creation, I would have given some useful hints for the better ordering of the universe."

He determined that the distance between Alexandria and Syene divided by the radius of Earth was related to the angle between the Sun's rays and the pillar—a calculation then at the frontier of mathematics, but today a high school geometry problem. (Fun fact: The very word "geometry" comes from the Greek word for "Earth measurement.")

Eratosthenes's result: The circumference of Earth is 50 times the distance between Alexandria and Syene, or 250,000 stadia.

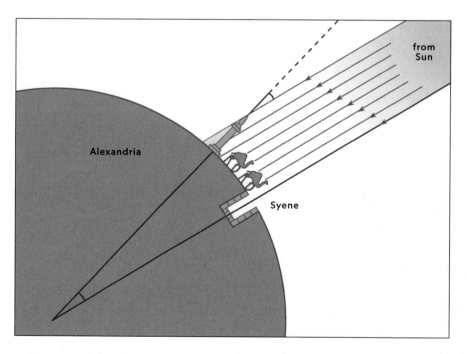

Comparing the angle of summer solstice sunlight shining down a well and at a pillar some distance away, Greek astronomer Eratosthenes observed the curvature of Earth—and estimated its circumference with remarkable accuracy.

The stade is a unit of length about a hundred yards long. (Our word "stadium" derives from this root.) Unfortunately, it wasn't a standard measurement—at least six different stadia were in use at the time—but given the most generous interpretation, his measurement was only about 10 percent off.

Not bad for a guy whose only instrument was a stick.

THE PARALLAX SOLUTION

To understand our place in the universe, we must ask how big the cosmos actually is, and how big it is compared to Earth. Sounds like a simple question. On Earth, calculating the distances to places and things normally poses little challenge. But in the universe, it opens a door to one of the most intricate problems in modern astrophysics—the distance ladder. We'll return to it several times before we're done.

Let's start by noting that our nighttime sky presents itself as a two-dimensional display of starlight. We know that those lights sit at different distances from us—or, in other words, we know the heavens are three-dimensional. The challenge is figuring out how far away everything is from Earth.

Unfortunately, the methods and tools that work for nearby objects fail for faraway objects. You have to stand on the first rung of the ladder, so to speak, to invoke the next set of methods and tools that take you farther. When those tactics reach their distance limit, you have to climb onto the next rung and find yet another method that works. And so on. Along the way, any distance uncertainties that came before worsen as you ascend the ladder.

The first rung in the cosmic distance ladder invokes parallax. You can demonstrate it for yourself with a procedure so simple, you've almost certainly done it before without knowing its implications. Hold your arm out and point your finger at an object across the room while you close your left eye. Now open your left eye and close your right. See how your finger appears to shift left and right against the background? That happens because the angle of sight from each eye to the fingertip is different. If you know those two angles and the distance between your eyes, some simple geometry gives you the distance from your eyes to the tip of your finger.

To extend this tactic to the sky, imagine measuring the angles of the lines of sight to a distant object like a planet from two different places on Earth. Again, knowing the angle from each point on Earth to the object and the distance between the observation points will give you the distance to the object.

The Greek astronomer Hipparchus applied this method to estimate the distance to the Moon and came up with about 60 times the radius of Earth: too large by a factor of two. (Considering

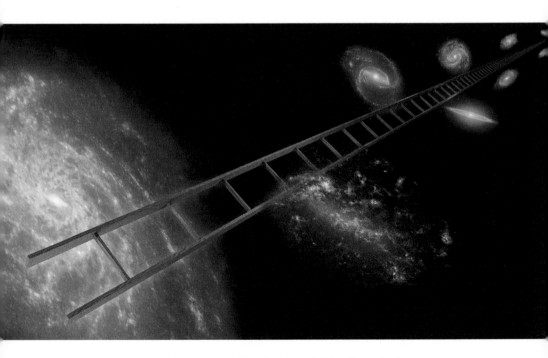

Exploring the cosmos is like climbing a ladder: From the first rung, you can observe certain things; with new tools and methods, you climb to the next rung and push farther into deep space.

MEET THE PARSEC

Angles are commonly measured in degrees, with 360 degrees composing a full circle. Each degree divides into 60 arcminutes, and each arcminute is divided into 60 arcseconds. For a star to show a parallax of one arcsecond, it would be located about 3.26 light-years from the Sun. This concept of one parallax-second has been shortened to the term "parsec"—a unit of distance that retains some currency in modern astrophysics and in space-based science fiction, including the *Star Trek* and *Star Wars* franchises .

he could have been off by a factor of 10 or 100 or 1,000, or wielded no method at all, you've got to admire the effort.) When he tried to get the distance to the Sun, he wasn't so successful. His calculations placed Earth closer than Mercury to the Sun.

But what happens when you want to measure the distance to really faraway objects, like the stars? Using the parallax experience with your pointer finger, the farther your hand is from your face, the less your finger shifts against the background as you wink your left and right eyes back and forth. If you had stretchy arms and held your pointer finger a football field's length away from your face, the winking would have little effect on the shift of your finger. The separation of your two eyes is small compared with the distance to your pointer finger, and so the angles rapidly become small and hard to measure.

Two solutions present themselves: (1) Develop instruments like the telescope that are better at measuring small angles, or (2) increase the separation of your eyeballs.

The telescopes would eventually arrive. And as parallax observations matured, the separation of our eyeballs—or two locations on Earth—would become the full diameter of Earth's orbit, as the process of parallax observation matured. Observe a "nearby" star against the background of stars much farther away. Wait six months while Earth journeys to the opposite side of its

orbit, and observe the star again. The shift you observe in the star's position in the sky is a cosmic version of winking your eyes. Now the baseline separation is not just eyeball inches but the full diameter of Earth's orbit. Yet even that baseline had to be measured by somebody.

HOW BIG IS THE SOLAR SYSTEM?

The universe as seen by medieval peasants was a small, homey place. Heaven was above their heads, and the stars and planets couldn't be much farther away than the next country. Even after Copernicus showed that the Sun, and not Earth, was the center of the known universe, it still seemed quite cozy.

But all that was about to change. In 1610, Galileo became the first person to turn a telescope toward the sky. As a result, he set in motion a chain of events that would eventually expand the universe in our minds to dimensions unimagined by the ancients. We think of the telescope as a groundbreaking instrument that allows us to see farther by magnifying images, but it also allowed astronomers to measure angles more accurately—which, in turn, enhanced our ability to measure small parallax angles and thus large distances.

In 1672, the nascent French Academy of Sciences sent an expedition to Cayenne in French Guinea to measure the position of the planet Mars on the sky, at the same time measurements were being made in Paris. The expedition was timed for a moment when Mars and Earth would be closest to each other, situated on the same side of the Sun. Using parallax and the known distance between the two telescopes, observers determined the distance to Mars. From this measurement, they used the laws of planetary motion Kepler worked out to calculate the distance between Earth and the Sun for the first time, dubbed the "astronomical unit" (AU). They came within 10 percent of the modern value.

A European Southern Observatory facility in Chile uses an artificial guide star beamed by laser to calibrate its VLTs—Very Large Telescopes—a modern way of coping with atmospheric turbulence.

30 CENTS Henrietta Leavitt's hourly wages at Harvard (about $9/hour today)

With this result, the universe suddenly became 20 times larger, and Earth more insignificant than ever before believed or imagined.

HENRIETTA LEAVITT & THE STANDARD CANDLE

With the latest spaceborne telescopes, parallax gets you the distances to the nearest billion stars. Sounds like a lot, but those stars occupy a tiny spherical volume around Earth, containing less than one percent of all the stars in the Milky Way galaxy. How can you measure the distance to a far-off star? Or to another galaxy? We need another rung on our distance ladder.

Enter Henrietta Leavitt, an extraordinary figure in the history of astrophysics. The daughter of a minister, she attended what was then called the Society for the Collegiate Instruction of Women, which later became Radcliffe College, the sister college to Harvard in Cambridge, Massachusetts. Upon graduation, she took a job at the Harvard College Observatory.

In those days the tedious analysis of astronomical data was done with pencil and paper and relegated to teams of people, usually women. They were called "computers." While analyzing various categories of stars, Leavitt made meticulous observations of a particular kind of rare star—a so-called Cepheid variable, named for the constellation Cepheus, in which the first of this variety of star was found. Leavitt discovered that its brightness waxes and wanes predictably over a period of weeks or months. She timed the variation and discovered that the longer it took a Cepheid variable to cycle, the more total energy the star emitted—that is, the brighter it was.

THE HARVARD COMPUTERS

In 1885 Henrietta Leavitt joined a team of women performing tedious measurements of stellar spectra. They were hired by Edward Pickering, director of the Harvard College Observatory, who reportedly wanted them to "work, and not think." Though highly educated, they were prohibited from operating telescopes and paid about as much as unskilled laborers. Upon Leavitt's discovery of the Cepheid variables, Pickering published her work under his name, and she never received the credit she deserved in her lifetime.

Henrietta Leavitt

If you know in advance the rate at which a star is emitting energy, then by using a simple formula, you can calculate how far away that star is just by measuring how bright or dim it happens to look at the distance you find it. But you first need a Cepheid variable close enough for you to establish its distance via parallax. Only then can you step to the next rung on the distance ladder. Leavitt's method was the first example of what astrophysicists call the "standard candle" technique for determining distance, which will surface again when we talk about the discovery of dark energy and the accelerating universe.

GALAXIES

By the early 20th century, astronomers had a pretty good grasp of Earth's place in our galaxy. Using Henrietta Leavitt's standard candle technique, the American astronomer Harlow Shapley established the size of the Milky Way: a whopping 100,000 light-years across. That measurement astonished astrophysicists of the day—and everybody else. The size of the universe was growing by leaps and bounds with every new measurement of distance. Shapley also established that the Sun is located not at the center of the Milky Way, but rather in the suburbs, two-thirds of the way out. This ego-busting discovery rivals that of Copernicus when he declared that Earth might not be the center of the known universe.

But wait, there's more.

Observers also noticed objects scattered across the sky that appeared as fuzzy shapes and blobs in their 1920s-era telescopes. Named "nebulae" (Latin for "clouds"), some of these objects were clearly glowing masses of amorphous gas and dust—and all of those happened to lie within the arcing band of light we call the Milky Way.

Another class of fuzzy objects, however, could be seen in all directions. Called spiral nebulae, these appeared as pinwheels on the sky: some edge-on, some at an angle, and some face-on. But

California's Mount Wilson Observatory empowered Edwin Hubble to confirm the existence of galaxies beyond our own. With its 100-inch, 4.5-ton mirror, carried by truck up a rocky mountain road in 1917, it was at the time the world's largest telescope.

$500,000   Approximate cost of the Mount Wilson 100-inch (2.5-meter) telescope—about $6.2 million today

the telescopes of the day could not distinguish individual stars within them.

At issue was the nature of the spiral nebulae. Were they, as Shapley asserted, simply structures within the Milky Way, like everything else in the night sky—implying that the Milky Way was in fact the entire universe? Or were they entire other galaxies, impossibly distant from us—veritable "island universes" scattered across the depths of space? Was the universe, in other words, a single massive collection of stars surrounded by emptiness, or was it composed of countless other galaxies, just like our own?

By the 1920s, the stage was set to answer this question. Philanthropist Andrew Carnegie helped fund the largest telescope of the day on Mount Wilson near Los Angeles, and a young man named Edwin Hubble was assigned to use it. (You guessed it: NASA's legendary Hubble Space Telescope is named in his honor.) The new powers of this telescope enabled Hubble to identify individual bright Cepheid variable stars in the spiral Andromeda Nebula, and he used Henrietta Leavitt's standard candle technique to determine their distance. The thing was more than two million light-years away—far too distant to be a nebula floating within the confines of the 100,000-light-year Milky Way. With these observations, Hubble revealed, once and for all, the large-scale structure of the universe: The Milky Way is simply one galaxy among many.

BILLIONS & BILLIONS

Having placed our planet in the solar system and the solar system in the Milky Way galaxy, we must now place our galaxy in the

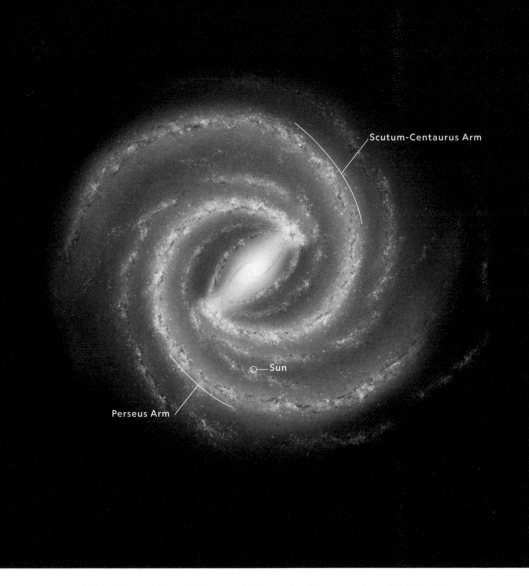

Scutum-Centaurus Arm

Sun

Perseus Arm

NASA's Spitzer Space Telescope helps us visualize our galaxy, the Milky Way,
now known to contain two major arms, Scutum-Centaurus and Perseus.
Our solar system lies on a spur between them, as shown in this illustration.

larger universe, possibly completing our quest to understand
Earth's place in the cosmos.

Once Hubble established the existence of galaxies, he
began a systematic survey and eventually developed a kind
of zoological classification scheme based on galactic shape.

It ranges from gas-empty, elliptically shaped galaxies that no longer make stars to gas-rich, grand-design spiral galaxies, like the Milky Way, where stars are born, live out their lives, and die (and where copious heavy elements are manufactured within stars and scattered across the galaxy). Most of these stars are likely to have planets circling them. Now add the planets wandering through space without stars, and you find we live in a galaxy with hundreds of billions of planets—some of which may support life.

Nevertheless, once we have accepted the ordinariness of our own galaxy, we are still left with discovering how many galaxies are in the universe and how far from us they happen to be—a problem not unlike what we faced when determining stellar distances within our own galaxy. For the most distant galaxies, we can no longer see individual stars, so Henrietta Leavitt's standard candle can no longer be used, forcing us to find another way to determine distance.

Within a few short years of Hubble's discovering that spiral nebulae were island universes—if that's not enough for you—he discovered that the galaxies were hurtling away from one another, with the more distant galaxies receding faster than the nearby ones. If the universe was smaller in the past than today, it may hint to a beginning of it all. But, in any case, this expansion of space stretches the wavelengths of light en route to us, causing a shift to the red of their spectral features—the now famous cosmological "redshift."

13,772,000,000

The age of the universe in years, plus or minus 59 million

So just measure a galaxy's redshift—a relatively easy task—and you get its distance from the Milky Way. Behold the next rung on the distance ladder. But, as before, you only get to step there after you've found a nearby receding galaxy that also harbors Cepheid variables visible within them.

Starting in the late 20th century, with ever more powerful telescopes, astrophysicists undertook extensive redshift surveys, which gave us three-dimensional maps of galaxies across the universe. The most exhaustive of these was the Sloan Digital Sky Survey, which cataloged the position in space of millions of galaxies.

Our best estimates put the number of galaxies in the observable universe at a hundred billion, and possibly two or three times that. There are, in other words, as many galaxies in the universe as there are stars in the Milky Way. And if each of these galaxies houses about the same number of stars as ours does, we're looking at more than 1,000,000,000,000,000,000,000 (one sextillion) stars in the observable universe.

A FINAL WORD

We have come a long way since Isaac Newton and Aristotle walked into that bar. Our views of our planet, ourselves, and our future have undergone stupefying changes, most of them related to our evolving view of the cosmos and our ever diminishing place within it: one indignity after the next.

People writing about this successive removal of humanity from the center of creation often throw in Charles Darwin, who taught us that we are really not all that different from other living creatures on our planet, and Sigmund Freud, who taught us that our mental processes aren't as rational and logical as we like to believe.

But there is one bright spot in this scientific dismantling of our ego. If there is nothing special about Earth—if we are really just part of a continuum of nature—then there's nothing special about the laws we discover on and from our own planet. These same laws likely operate everywhere else, empowering us to explore and decode the entirety of the known universe, across space, and possibly across time itself. There's no denying it. What's bad for our ego is good for science.

Composed of 800 exposures taken over 11 days in 2003–2004 by the Hubble Space Telescope, this Ultra Deep Field image captures nearly 10,000 galaxies. The deepest space image ever made, it includes galaxies dating back 13 billion years.

HOW DO WE K
WE KNOW?

Time exposure of lights and stars at the La Silla Observatory in Chile

NOW WHAT

2

Have you ever ventured far away from obtrusive city lights and looked up on a clear night? Did you feel awash with awe at the brilliant tapestry of stars?

Long ago, that's what everybody saw, every night. Even in bustling cities. For our ancestors, every night was like that experience you had when you were away from city lights. A stately progression of the stars and planets was simply part of their lives. For this reason, astronomy may well be humanity's first science, if not the world's second oldest profession.

But all that changed with the introduction of street lighting in the 19th century, when starry nights slowly faded from our view, lost to an ever brightening urban sky glow, punctuated only by the Moon, some planets, and the brightest of stars.

The academic discipline called archaeoastronomy provides our best evidence for the antiquity of astronomy. Only a few decades old, this field studies cultural artifacts—especially structures—to determine what ancient civilizations knew about astronomy and what role that knowledge played in their lives.

Greek astronomer Hipparchus compiled the first star catalog,
in the second century B.C.

The best known example of this sort of study concerns the famous circle of massive stone pillars called Stonehenge, on England's Salisbury Plain.

For the record, Stonehenge was not built by the Druids, nor Julius Caesar, nor Merlin the magician teleporting the stones in from Ireland, nor extraterrestrials in flying saucers. We now know that it was built by a series of peoples, beginning in about 3000 B.C. and ending some 1,200 years later. None of these people had a written language or the wheel and axle, but over the centuries they produced the monument still standing today.

Stone circles such as Stonehenge, likely designed to track the Sun and the seasons, show prehistoric cosmic awareness.

Just because you can't figure out how ancient civilizations built stuff, doesn't mean they got help from Aliens.

British-American archaeoastronomer Gerald Hawkins was the first to suggest that the layout of Stonehenge might have cosmic significance. He grew up near the monument, in the time before it was fenced off. Playing among the stones, he noticed clear lines of sight defined by their placement—almost as if those ancient builders wanted to compel visitors to "Look this way." Later, while at MIT, Hawkins had access to modern 1970s-era computers and showed that many of the lines of sight at Stonehenge pointed to significant astronomical events. The most famous of these marks the rising of the Sun on the summer solstice. Thus, one function of Stonehenge was to keep track of the seasons: a vital skill for an agricultural society.

Since this discovery, ancient structures all over the world have revealed similar results. Perhaps the most striking are the medicine wheels found in western North America, built not by

CALENDARS OF STONE

Hundreds of ancient stone medicine wheels dot the landscapes of North America, constructed by nomadic Native Americans including the Sioux, Cheyenne, Crow, Blackfoot, Arapaho, Cree, Shoshone, Comanche, and Pawnee. The most famous of these, the Big Horn Medicine Wheel in Wyoming, has a radius the length of a Greyhound bus, with 28 spokes connected to a central cairn. This wheel and others like it were probably used to predict the positions of the Sun and bright stars in the night sky across the seasons.

Tycho Brahe with part
of his nose

TYCHO'S NOSE

At a drunken student party, Tycho got into an argument with another student about which of them was a better mathematician. This led to a duel, and Tycho wound up with the tip of his nose cut off. In astronomy folklore, Tycho wore a nose prosthesis made of gold and silver. When his body was exhumed in 2010, primarily to investigate the cause of his mysterious death, scientists put that claim to rest. Chemical analysis of his nasal bones showed traces of copper and zinc: In other words, his false nose was brass.

farmers, but by nomadic peoples. Such relics show that the skies provided critically important cues to our ancestors, no matter where they lived or how they sustained themselves.

NAKED EYE ASTRONOMY

Astronomy developed without telescopes. This simple and obvious statement carries deep implications. Ancient astronomers knew that Earth is a sphere. In fact, even without telescopes, they compiled an impressive list of achievements. As we've already seen, around 100 B.C., Hipparchus of Nicaea measured the position of prominent stars and made a credible measurement of the distance from Earth to the Moon. He also discovered the precession of the equinoxes, a very slight wobble in Earth's axis of rotation. Meanwhile, also without telescopes, Aristarchus of Samos developed a heliocentric model of the solar system. And Claudius Ptolemy, around A.D. 150, produced a complex

NASA's Wide-field Infrared Survey Explorer reveals the remnant of Tycho Brahe's 1572 supernova (in red) that he and his contemporaries saw.

TYCHO & SHAKESPEARE

Tycho's nova is actually mentioned in the first act of Shakespeare's *Hamlet*, written around 1600. Bernardo, a guard at Elsinore Castle, speaks of "yond same star that's westward from the pole" and that has "made his course to illume that part of heaven / Where now it burns." People at the time would have recognized this quote as referring to the new star that Tycho Brahe had studied.

A supernova seen in the constellation Cassiopeia so impressed Shakespeare that he mentioned it in Hamlet.

model of the solar system that dominated astronomical thinking for nearly 1,400 years.

The basic tool of the naked eye astronomer is nothing more than a sighting tube aimed at a star or planet. You then determine the object's position on the sky with two numbers—the angle above the horizon and the angle away from a compass direction already established and agreed upon, such as due north. Therein lies the foundation of celestial navigation. When you make measurements this way, size matters. The longer the tube, the more accurately you can determine the position of the object in the sky.

The king of the naked eye astronomers was Tycho Brahe. Born into a prominent Danish family, he quickly established himself as a major force in European astronomy. In his 20s, he studied the nova of 1572. Observers on Earth saw a star in the sky where no star had been seen before—an anomaly to all, since the Bible taught that stars in the heavens were fixed and unchanging. Of course, *nova* is Latin for "new." Tycho's careful observations showed that the nova, now known to have been a supernova, was not an atmospheric phenomenon, but must sit farther away from Earth than the Moon.

The Danish king, thrilled that a member of his court had become famous, gave Tycho the entire island of Hven and some cash to build Uraniborg Observatory—the world's first federally funded research institute. There, Tycho built huge state-of-the-art sighting tubes along with supporting instruments and compiled the most accurate data on planetary motion ever before made.

GALILEO & THE TELESCOPE

Sometimes a single event is responsible for such critical and far-reaching consequences that its import can only be

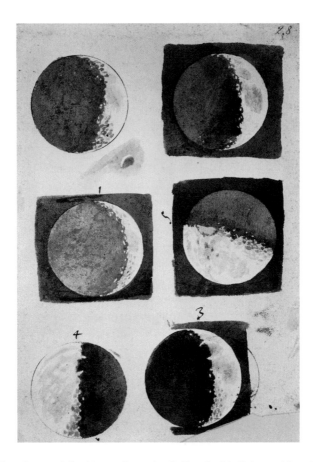

The phases of the Moon, drawn by Galileo for his *Sidereus Nuncius*

appreciated from a distance. We suggest that a night in 1610, when the Italian astronomer Galileo Galilei first turned a telescope toward the sky, was just such a moment. It forever changed the way that human beings regard the cosmos.

Galileo didn't actually invent the telescope; as it happened, word of a new Dutch "spyglass" percolated throughout Europe in the first decade of the 17th century. Upon hearing of the new device, Galileo immediately improved the existing design, better suiting it for astronomical applications. Still, the magnification power of Galileo's telescope would be matched by a moderately priced pair of binoculars today.

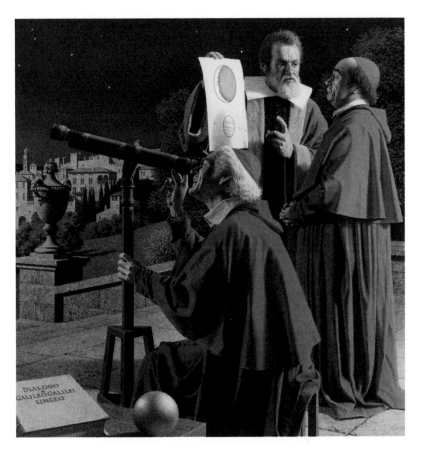

With a telescope, Galileo could see features on the Moon's surface,
yet many contemporaries remained skeptical of his observations.

What Galileo saw rapidly forced the classical picture of the
solar system, developed by thinkers like Aristotle and Claudius
Ptolemy, to be abandoned. Conventional religious and astronomical wisdom in 1610 held that Earth was the unmoving
center of the universe, above which the heavens reigned as pure,
perfect, and unchanging. Overturning conventional scientific
wisdom can and will make you famous. Overturning conventional religious wisdom can land you in front of the Inquisition.
Galileo did both.

So you can imagine Galileo's surprise when he saw:

1835 The year the Catholic Church removed Galileo's books from the index of forbidden literature

◼ **Surface Features of the Moon** | Like all the bodies in the heavens, the Moon was supposed to be a perfectly smooth sphere.

◼ **Sunspots** | Like the Moon, the Sun was supposed to be perfect, without blemish. Galileo saw what we would call today a sunspot group.

◼ **Phases of Venus** | Everything in the heavens was supposed to be in orbit around Earth, but Venus could only have phases—changes from crescent to half to gibbous and back, like the Moon—if both it and Earth are orbiting the Sun.

◼ **Moons of Jupiter** | Earth was at the center of everything—the home of humanity, the crown of creation—yet here were four celestial bodies that seemed perfectly happy to be in orbit around Jupiter.

Taken together, these observations were powerful evidence against the classical picture of the geocentric universe and for the heliocentric model, published 65 years earlier by Nicolaus

NAMING THE MOONS

In *Sidereus Nuncius*, Galileo named the four largest and brightest satellites of Jupiter the Medicean moons in an attempt to flatter Duke Cosimo de' Medici of Florence. The ruse apparently worked, because Galileo received a large cash award. Today, more fittingly, these four—Io, Europa, Ganymede, and Callisto—are called the Galilean moons.

1992 The year Pope John Paul II and the Vatican's Pontifical Academy of Sciences decided that Galileo was right all along

Galileo viewed some of Jupiter's moons through his telescope.
Today, thanks to NASA's *Galileo* mission, we see them in greater detail,
as displayed in this composite image of Io.

GALILEO'S TRIAL

The history of Galileo's conflict with conventional theology is long and complex. He was warned not to defend the Copernican system in his writing. The fact that in his *Dialogue Concerning the Two Chief World Systems* he put the Pope's favorite arguments in the mouth of a character named Simplicio ("Fool") probably didn't help his case when he was tried on suspicion of heresy. One famous legend claims that after Galileo was forced to confess he had never believed what he actually wrote or said, he muttered in a low voice, speaking of Earth: *"Eppur si muove*—and yet it moves."

Copernicus. Galileo reported his findings in a tiny book titled *Sidereus Nuncius (Starry Messenger)*. He wrote well and, more importantly, he wrote his later books in Italian, not Latin. This meant his ideas were available to the literate strata of Italian society, and not just to scholars. Galileo's revolutionary ideas armed his enemies with reasons for attacking him and, eventually, putting him on trial for suspicion of heresy—a charge of which he was found not guilty only when he recanted his statements.

THE ELECTROMAGNETIC SPECTRUM

Until now we have focused on knowledge brought to Earth in the form of visible light. There are many good reasons that this was the way we came to know what we knew in centuries past.

First and foremost, we are primates, and like our primate kin, we sense the world primarily through sight. So cherished is this mode of knowing that when we say "I see," we mean "I understand." Naturally, the first instruments to probe the heavens were our eyes.

Second, Earth's atmosphere is transparent to visible light. If you've ever seen the lights of a distant city from an airplane at night, you know that light can travel miles through the bottom

Many insects see ultraviolet light; others see light in the infrared portion of the electromagnetic spectrum. For them, flowers look like this.

of our atmosphere unimpeded. And the best evidence that the top of our atmosphere is transparent to visible light is that you can see the Sun, Moon, and stars in the daytime. A simple but profound experiment.

In addition, the Sun, our primary source of illumination, has a surface temperature of about 5000°C, with a peak energy output in the form of visible light. As daytime creatures, we should not be surprised that natural selection favored a sensory organ—the human eye—that is exquisitely sensitive to this particular form of energy.

In contrast, imagine the view from the surface of Venus, where the entire planet is shrouded in thick clouds 24/7. A trickle of diffuse visible light gets through in the daytime. You would, of course, vaporize in the furnace temperatures caused by a runaway greenhouse effect. But ignoring that complication, humans evolving on that world would know nothing of a night sky, delaying the advances of astronomy by millennia, or perhaps even preventing the science from arising at all.

Light is a form of electromagnetic radiation or electromagnetic waves. All the colors of the rainbow correspond to different wavelengths of the wave. Red light, for example, has a wavelength roughly 8,000 atoms across. Violet light has a wavelength roughly half that. When the Scottish physicist James Clerk Maxwell codified the laws of electricity and magnetism in the late 19th century, his equations predicted not only the existence of electromagnetic waves, but also that they must exist at all wavelengths—in both longer and shorter wavelengths far beyond what we can see. Why, then, are humans only aware of "visible" light? Our paltry ability to perceive but a tiny sliver of an infinite spectrum is like attending a full symphony orchestra but hearing only the piccolo.

Not long after Maxwell's predictions, the German physicist Heinrich Hertz discovered electromagnetic wavelengths running from a few feet to many miles. You know them now as radio waves. These were the first of an ensemble of electromagnetic waves to be discovered that live entirely outside our sensory experience. In fact, Maxwell was quite correct in predicting the existence of waves at all wavelengths.

Turns out that except for radio waves and visible light, Earth's atmosphere blocks all other forms of electromagnetic radiation. After crossing light-years of space, that radiation is absorbed in the last few miles of its journey. The universe has always been sending us information in every part of the electromagnetic spectrum—and yet we were clueless.

THE RADIO UNIVERSE

Earth's atmosphere is transparent to radio waves, as well as to visible light. You can see this for yourself: You can make and receive cell phone calls in practically any indoor environment. This is only possible because radio waves, in this case short versions of the radio waves called microwaves, have traveled from a tower to your smartphone, which converts the signal into the sound of your phone call.

Unfortunately, the human body is not equipped with "radio eyeballs" that can detect these radio waves the way we can detect visible light. Consequently, the field of radio astronomy did not exist until James Clerk Maxwell and Heinrich Hertz called this form of electromagnetic radiation to our attention. You can think of the atmosphere as having two "windows": one for light, another for radio. For all the rest of the electromagnetic spectrum, the atmosphere might as well be a brick wall.

But we can learn a lot by peering through the radio window. The first instrument designed to detect radio waves from space—the first radio telescope—was built by the American engineer Karl Jansky at the Bell Telephone Laboratories in New Jersey in 1932. Jansky was tasked with finding radio signals coming from the sky that might interfere with radio communications on Earth—but instead wound up discovering radio signals coming from the Milky Way. In 1937, American radio engineer Grote Reber, fascinated by Jansky's discovery, built his own telescope specifically for studying these galactic radio emissions in the backyard of his home in Wheaton, Illinois. His surveys mark the birth of radio astronomy.

Radio waves are the weakest on the electromagnetic spectrum, so radio telescopes are often enormous structures designed to collect the maximum amount of radiation. They can be large steerable dishes, but the biggest radio telescopes are nonsteerable

Karl Jansky's antenna swiveled on Model T tires—some called it his merry-go-round. With it, he was the first to detect radio waves emanating from space.

and built into depressions in the ground. They observe what drifts past their wide fields of view as Earth rotates. The largest such telescope is the Five-hundred-meter Aperture Spherical Telescope (FAST), completed in 2016 by China. The dish is so large, it could fit four 100,000-seat football stadiums.

Over the decades, radio telescopes have empowered astrophysicists to detect objects in the universe that emit primarily radio waves. One point cannot be overstated—these objects would otherwise remain invisible to us. For example, radio telescopes discovered pulsars—rapidly spinning, compact cores of stars left over by supernova explosions. The radio signal emanates from one region, but the spin, like a lighthouse sweeping across the horizon, produces regular pulses detected by the telescope. Hence, the name pulsar. And hence, the brief but

---

ET, USE RADIO WAVES

Most of the searches for signals from extraterrestrials have been conducted with radio telescopes. Why would aliens communicate with radio waves instead of any other wave band on our electromagnetic spectrum? First of all, radio is the most energy-efficient frequency, requiring the least amount of energy to generate. Gamma rays, at the other end of the electromagnetic spectrum, carry the most energy and therefore require the most energy to transmit. Additionally, there is far less interference from other celestial phenomena at selected radio frequencies. Radio waves pass through obscuring gas and dust clouds in the galaxy as though those obstructions aren't there at all. So if aliens really want to send signals through the universe, and they have astrophysicists among them who understand the universe as we do, they would most likely transmit their messages with radio waves.

---

initial suspicion that aliens were at play, sending repeated radio pulses for us to decode.

Whenever we open new windows to the sky, unexpected phenomena like these remind us of how much we don't know about what we don't know.

FROM ASTRONOMY TO ASTROPHYSICS

Sciences, like people, undergo phases as they mature. What we know, and how we know it, changes at the same time. Until the 19th century, for example, biologists primarily concerned themselves with cataloging the species of life on Earth; today, on the other hand, they care about the molecular and physical processes that govern life. Biology, in other words, has absorbed chemistry and physics into its ranks.

Similarly, up to the mid-19th century, classical astronomy focused on the brightness, color, and location of objects in the sky—so much so that French philosopher Auguste Comte formally declared these bits of knowledge as the fundamental limits of astronomy. "Any research which is not ultimately reducible to simple visual observations is . . . prohibited for us concerning the stars," he wrote in his *Cours de Philosophie Positive* of 1835, which meant that we "could never study by any means their chemical composition, or their mineralogical structure."

That's surely one of the most boneheaded statements ever made by a learned person. What followed, just a few decades later, were discoveries about every aspect of the stars that Comte declared unknowable: chemical composition, density, temperature. This change was driven by a new branch of chemistry and physics called spectroscopy, seamlessly adopted into astronomy, heralding the birth of modern astrophysics.

FAST, the Five-hundred-meter Aperture Spherical Radio Telescope in China, is nicknamed Tianyan—"eye of heaven."

Neil deGrasse Tyson ✔
@neiltyson

The largest Telescope in the world, a mile in circumference, is no longer in the USA. It's in the Guizhou province of China. So when Aliens say "Hi", the first humans to receive their signal will be Chinese Astrophysicists.

💬 1.6K   ↻ 6.1K   ♡ 27.6K   3:42 PM · Aug 3, 2018

Two scientists pioneered this new field at the Heidelberg University in Germany—an institution that, until the rise of Adolf Hitler, was one of the most prestigious scientific centers of learning in the world. The first was Robert Bunsen, a chemist and, as you surely suspected, the inventor of the Bunsen burner, which you may remember from your high school chemistry lab. His studies of gases emitted by the production of cast iron propelled Germany to dominance in the heavy metal industry. The second was Gustav Kirchhoff, a physicist whose laws governing electrical circuits are still learned by university students throughout the world.

Bunsen studied the light emitted when various elements were heated, and Kirchhoff suggested passing the light through a prism. A prism breaks white light, or any light, into its component colors, bending them each through different angles, thus generating a spectrum. When this happens with raindrops in sunlight, you get a rainbow.

The two researchers found that pure chemical elements, when heated, gave off a distinctive and characteristic set of features that appear as lines in the spectrum—and that the pattern of these lines was different from one element to the next. The spectrum of an element was a kind of fingerprint that could be used to detect the presence of that element in whatever was being heated.

THE FIRST SPECTROGRAPH

Bunsen and Kirchhoff built the first spectrograph—an instrument for measuring spectra—from a couple of old surveyors' telescopes, a prism, and, believe it or not, a cigar box.

So, how does this relate to astrophysics and the ways we have come to know what we now know? It doesn't matter how far away the source of light is; its signature spectrum is always the same. It can be at the other end of a laboratory bench or in a galaxy billions of light-years away. Once the light is emitted, it carries a distinct spectrum, containing multiple fingerprints that identify the nature of the source. This fact empowers astrophysicists to identify the chemical content of stars and exoplanets, allowing us to pay a little less attention to "Where is it?" and more attention to "What is it?"

KNOWLEDGE FROM ABOVE THE ATMOSPHERE

We have seen that the atmosphere is transparent to radio waves and visible light. But the visible light window isn't perfect. The turbulent movement of air as light passes through it blurs the image we receive on the ground—and incidentally, also makes stars twinkle. So don't ever wish an astrophysicist a night of twinkling stars.

How to transcend these limitations? One obvious solution is to put our receivers above the atmosphere, rather than on the ground below it.

Another strategy, appropriate only for objects you can get to, is to send a spacecraft to the source of the radiation and get a close-up view. At the moment, that works for places within the solar system. Beyond that, we've got work to do.

IN LOW EARTH ORBIT | Of all the destinations in the solar system, the easiest place to send a satellite is into Earth orbit. It requires less energy to put a satellite into orbit than to send it anywhere else. Also, if the altitude is in range of our crewed

1,000,000 The number of miles from Earth where we find Lagrange points L1 and L2

---

vehicles, then astronauts can visit to make repairs and install updates, allowing the observatory in question to function for a long time (as was done with the Hubble Space Telescope). Earth orbits are also useful for global positioning system (GPS) satellites, as well as satellites designed to monitor Earth systems, including sea level and temperature.

AT LAGRANGE POINTS | Named for the Italian mathematician Joseph-Louis Lagrange, Lagrange points are locations in space where all forces of motion and gravity are in balance between two cosmic objects—Earth and the Sun, or Earth and the Moon, for example. Anything placed there will have no tendency to fall in one direction or another.

Newton's first law of motion tells us that any object sent into space will either continue moving in the direction we sent it or surrender to the gravity of another object. At Lagrange points, an object rests between an equal push and pull of all forces. They are veritable parking spots for space hardware.

Any two-body system has five Lagrange points. The one labeled L1, located between Earth and the Sun, provides NASA's and the European Space Agency's solar telescope an uninterrupted view of the Sun. L2, located on the far side of the Earth–Sun system, allows an uninterrupted view of deep space and is the future location of the James Webb Space Telescope.

VIA SPACECRAFT | We have devised many ways to gain knowledge from here on Earth, but the best way to get to know a celestial object is to visit it. Humans have launched a flotilla of spacecraft over the past decades to explore the rest of the solar system. By now, spacecraft have flown by every planet in our system, orbited a few, and investigated a number of asteroids and comets. Some significant milestones:

■ **The twin Voyager spacecrafts,** launched in 1977, were the first artificial objects to leave the solar system, Voyager 1 in 2012 and Voyager 2 in 2018.

■ **The Galileo spacecraft,** launched in 1989, discovered the subsurface oceans of Europa while orbiting Jupiter from 1995 to 2003, thereafter plunging to its death in Jupiter's atmosphere.

■ **The New Horizons spacecraft,** launched in 2006, flew by Pluto in 2015, as well as a Kuiper belt object, 486958 Arrokoth, in 2019, as it continues to exit the solar system.

■ **The Cassini spacecraft,** launched in 1997, arrived at Saturn in 2004. It remained in orbit there for 13 years, providing stunning images and unprecedented data on the planet, its ring system, and its moons. Cassini further unveiled worlds like icy Enceladus, where, as on Jupiter's Europa, extraterrestrial life may thrive in subsurface oceans. Mimicking the Galileo spacecraft, in 2017 Cassini made its final death-dive through Saturn's atmosphere, ending its mission.

OPENING NEW WINDOWS ON THE UNIVERSE

Electromagnetic waves aren't the only source of information from the heavens. Other kinds of waves and an entire rogue's gallery of particles also reach Earth. Each time we learn how to

Neil deGrasse Tyson ✓
@neiltyson

Farewell Cassini, how far you've come. On this eve, in fiery death, Saturn & you are one. VIP (Vaporize In Peace): 2004-2017

SEPTEMBER 15, 2017
END OF MISSION

💬 795   🔁 31.2K   ♡ 94.6K   9:51 PM · Sep 14, 2017

detect and understand these visitors, we open a new window on the universe. Two fresh windows recently opened: neutrinos and gravitational waves. And more await our clever technologies to open them, including dark energy and dark matter.

NEUTRINOS | The neutrino (meaning "little neutral one") has no electrical charge and almost no mass. A fundamental particle, neutrinos are copiously produced in nuclear reactions and hardly ever interact with matter. As you read this, one hundred billion of them pass through every square centimeter of your body every second, but only a few will ever jostle even one of your atoms in your lifetime.

The only way to detect neutrinos, then, is to force-feed them lots of atoms with which to interact. This is the idea behind IceCube, a giant neutrino detector located at the South Pole.

Hot water bores holes in the ice, into which cables carrying light detectors are lowered. Then the water freezes around them. When neutrinos jostle an atom in the ice, these detectors see a characteristic flash of light. By this clever technique, IceCube transforms an entire cubic kilometer of Antarctic ice into a dedicated neutrino detector.

Even more amazing, some of the neutrinos IceCube detected will have hit Earth at the North Pole and traveled all the way through the planet without interacting with a single atom before they enter the cubic kilometer of ice at the South Pole.

GRAVITATIONAL WAVES | The general theory of relativity predicts that if massive objects are accelerated, they will emit a wave in the space-time continuum. Picture them as outgoing

Two LIGO observatories—this one in Louisiana, another in Washington State— collect, compare, and confirm evidence of gravitational waves.

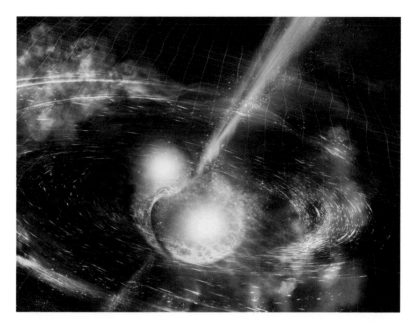

An illustration depicts the collision of two neutron stars, a cosmic event whose disturbance in the fabric of space and time can be detected by LIGO.

ripples of water on a smooth pond. These waves are very weak, but they have a characteristic influence on the material objects they encounter. In a comically exaggerated image, if we start with a basketball, then a gravitational wave will squash it into a football shape and then back to a basketball as it passes through. In reality, these distortions are tiny—smaller than the diameter of particles within the nucleus of an atom.

Enter LIGO (Laser Interferometer Gravitational-Wave Observatory), a research facility equipped with two arms in the shape of an L, each four kilometers long. A laser shoots light down each arm to a mirror, which reflects it back. If a gravitational wave washes over it, the path length will briefly change, shifting the mirrors relative to each other.

On September 14, 2015, LIGO recorded the first measurement ever of a gravitational wave, a century after Einstein predicted them. The wave, generated by the collision of two black holes

DUPLICATING EFFORTS

There are actually two detectors in the original LIGO array—one in Louisiana and one in Washington State. This two-facility system was built to avoid accidental events (or intentional pranks) masquerading as gravitational waves at one location or the other.

with masses 36 and 29 times the mass of the Sun, emanated from a long time ago in a galaxy far, far away: about 1.5 billion light-years from Earth. Since then, other countries have built gravitational wave detectors, developing what will ultimately become a global network.

OBSERVATORIES TODAY

Hundreds of astronomical observatories are scattered over Earth's surface, and we have dozens more in space, each one with its own role in contributing to our knowledge. What follows are some of the most spectacular of these windows to the cosmos.

OBSERVATORIES ON THE GROUND I The main problem with observatories on the ground is the imperfect, fuzzy view they offer through our atmosphere and out into the cosmos. Air turbulence and the presence of disruptive gases such as water vapor can distort images. To minimize this problem, the world's leading observatories are typically located at high altitude, above most of the weather that confounds observations. The two highest:

■ **Mauna Kea Observatories** I Located more than 14,000 feet (4,200 m) above sea level on the Big Island of Hawaii, the Mauna Kea Observatories house more than a dozen different telescopes. The site sees all of the northern and

ALMA—the Atacama Large Millimeter/submillimeter Array in Chile—combines 66 antennas in a desert location more than three miles above sea level.

most of the southern sky and is located in the middle of the Pacific Ocean, offering a stable atmosphere and a minimum of twinkling. Assorted detectors mounted on these telescopes monitor electromagnetic radiation from infrared to visible light as well as microwave radiation.

Atacama Observatories | Other than parts of Antarctica, the Atacama Desert in Chile is the driest spot on the planet. Some areas have no recorded rainfall, ever. Add to this an elevation above 15,000 feet (4,570 m), and we've got one of the choicest spots in the Southern Hemisphere to locate a telescope. One important installation is the Atacama Large Millimeter/submillimeter Array (ALMA), radio telescopes that, because of their elevation, capture microwaves before they are absorbed by water vapor in the lower atmosphere.

OBSERVATORIES IN SPACE | The Hubble Space Telescope, launched in 1990, remains the crown jewel of space observatories and is arguably the most productive telescope ever built. In fact, based on the sheer number of research papers and international collaborators engaged, Hubble may be the most productive scientific instrument of any kind, ever. Unfortunately, Hubble now nears the end of its productive life, and NASA has no further plans for service missions.

The Hubble is one of several "Great Observatories" in NASA's portfolio. The others are the Compton Gamma Ray Observatory, the Chandra X-ray Observatory, and the Spitzer infrared space telescope. The observatories are named for scientists Arthur Compton, Subrahmanyan Chandrasekhar, and Lyman Spitzer. Together with the ground-based observatories already described, they provide astrophysicists with information from every known part of the electromagnetic spectrum.

The huge mirror of the James Webb Space Telescope, more than six times bigger than the Hubble's, is made of 18 hexagonal units designed to unfold once the spacecraft reaches its Earth–Sun Lagrange point.

In addition, a wide variety of probes and specialized observatories operate in deep space. More than 40 NASA missions are active today, and other countries have followed. Recently, China landed a probe on the far side of the Moon, and the European Space Agency (ESA) launched a state-of-the-science observatory to study exoplanets.

COMING ATTRACTIONS

At any moment, dozens of scientific proposals are afloat, each with an innovative idea that aims to solve an important problem. Unfortunately, they cannot all be funded. But here are a few that have run the gauntlet of peer review, funding, and development, and will soon be in operation—and another that still has a way to go.

JAMES WEBB SPACE TELESCOPE ❘ Designed as the successor to the Hubble Space Telescope, the James Webb Space Telescope's 21-foot (6.5-m) mirror, significantly larger than Hubble's, is constructed from 18 interlocking hexagonal segments made of gold-coated beryllium. It will observe wavelengths from visible light into the mid-infrared. Its detection abilities will be unmatched for a spaceborne telescope, with a primary goal to detect objects with extremely large redshifts—the oldest objects in the universe. These are nascent galaxies with light that began in the blue part of the spectrum but got redshifted by the expanding universe as it traveled, all the way to the infrared by the time it reaches us. After launch, the James Webb telescope will park itself at the Earth-Sun L2 Lagrange point, shadowed from the Sun by a large shield, and unfurl its mirror segments, which were cleverly folded for the journey there. Because this point is a million miles from Earth, there is no chance that astronauts can make repairs and upgrades as they do for the Hubble. So the James Webb Space Telescope absolutely must work on the first and only attempt at deployment.

EXTREMELY LARGE TELESCOPE | This aptly named instrument is funded by the European Space Agency and is under construction in the Atacama Desert of Chile. Its primary mirror is a staggering 130 feet (40 m) across. For comparison, the diameter of the largest visible-light telescope built in the 20th century was one-fourth that diameter. In addition to its size, the Extremely Large Telescope (ELT) will use a technology called adaptive optics to produce sharp images more than 10 times as detailed as Hubble. To accomplish this, the various segments of the mirror will be deformed in real time to compensate for changes in the atmosphere. This will allow the ELT to examine and obtain images of exoplanets—and perhaps to detect water and organic compounds in protoplanetary disks that will someday evolve into planetary systems.

LASER INTERFEROMETER SPACE ANTENNA | This system represents the next step in detecting gravitational waves. Also funded by the European Space Agency, the Laser Interferometer Space Antenna (LISA) will consist of three free-flying satellites located at the corners of an equilateral triangle whose sides are more than six times longer than the distance between Earth and the Moon. The triangle will orbit the Sun but trail Earth by roughly 30 million miles (50 million km). Like its earthbound predecessor, LIGO, though much more sensitive, LISA is designed to detect the ripples in space-time associated with gravitational waves, in this case by tracking the relative positions of test masses located in each of the three satellites. LISA represents the next step in opening the gravitational window into the heavens.

Set to start up in 2025, the Extremely Large Telescope (shown here as an artist's concept) joins other observatories in the high, dry Atacama Desert, where water vapor and turbulence in the atmosphere are least likely to obscure the view.

HOW DID THE
GET TO BE TH

UNIVERSE
IS WAY?

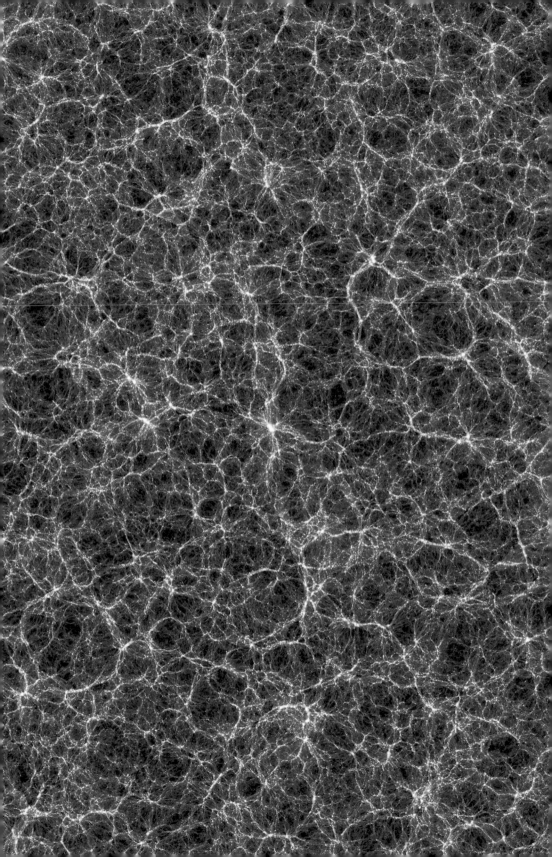

3

Edwin Hubble didn't just establish the existence of galaxies. He also discovered the universe's most salient feature: It is expanding. Hubble used the standard candle method developed by Henrietta Leavitt to determine the distance to nearby galaxies, and then created a whole new rung on the distance ladder by combining these measurements with a telltale feature of distant galaxies: that the wavelengths of light emitted by atoms and molecules in those galaxies are measurably longer than they would be when measured in the laboratory. We call this phenomenon redshift, because red light has the longest wavelength of all the colors visible to the human eye. It was a vital discovery that would continually serve our quest to learn about how the universe got to be the way it is today.

Here's a profound fact of physics. Objects that exhibit redshifts are moving away from you— and the bigger the shift, the faster they are moving. When Hubble sorted the redshifts of

A supercomputer in Germany took a month to generate
three-dimensional simulations of dark matter in the universe,
resulting in images including this one.

galaxies by distance, a mind-blowing fact sprung forth. The farther away a galaxy is, the faster it is moving away from us. This is known as Hubble's law and can be written in equation form as:

$$v = Hd$$

where v is the velocity of the galaxy, d is the distance to it, and H is a number known as Hubble's constant. In other words, the universe is expanding.

Rather than think of galaxies moving outward like a fireworks display, instead picture raisins scattered throughout rising bread dough. Now imagine hanging out on one of those raisins. You would see all the other raisins moving away from you. And the farther away from you another raisin was, the faster it would be moving, simply because more expanding bread dough exists between you and it. The raisins aren't moving through the dough, but rather they are carried along by the expansion of the dough itself.

One final point about our bread dough universe: Any raisin you choose to stand on offers the same view as every other raisin. You will perceive your own raisin as stationary, while all other raisins appear to be moving away from you. Every raisin sees itself as the center of an expanding dough universe. In the accidentally prescient words of the medieval philosopher Nicholas of Cusa, "The universe has its center everywhere and its edge nowhere."

The model of our expanding and cooling universe beginning at a specific time in the past, when all matter and energy were in the same place at the same time, is called the Big Bang, and it contains the core ideas of how the universe evolved into what we know it to be today. And it turns out that modern analysis of the expansion allows us to estimate the age of the universe with exquisite accuracy.

THE BIG BANG

Generally, when materials are compressed, their temperature rises. If you have ever inflated a bicycle tire using a hand pump, you may have noticed the valve warms up as your actions continue—a direct result of compressing the air within the pump cylinder.

The universe operates similarly. If you imagine running the film of the Hubble expansion backward, the universe would get hotter as we got smaller.

Imagine that you have confined steam at a high temperature and pressure, like inside a rice cooker. What happens when you release the pressure? The steam expands and cools. It will, in fact, keep cooling until its temperature reaches 212°F (100°C). At this

---

THE DOPPLER EFFECT

Nineteenth-century Austrian physicist Christian Doppler discovered his celebrated principle by studying the changing pitch of a train whistle as the train moves by. If the source of a wave—be it light or sound—is in motion, different observers will perceive that wave differently, depending on whether the source is moving toward or away from them. If the source of a sound wave moves toward you, the distance between the crests of the wave—the wavelength—will be shorter than if the source remained stationary, giving you a higher pitch. And if the source moves away from you, the wavelength will be longer, generating a lower pitch. For light waves, a receding source produces waves that have longer (redder) wavelengths.

You may have had more experience with the Doppler effect than you thought. You just need to pay attention. Next time a speeding ambulance approaches and then recedes from you, notice the change in pitch of its siren, becoming higher as it approaches, then dropping down lower as it moves away.

Neil deGrasse Tyson ✔
@neiltyson

A reminder that in a baseball game you cannot blame an Umpire's enlarged strike zone on the expanding universe.

💬 280   ↻ 3K   ♡ 11.3K   7:20 PM - Oct 21, 2016

moment, an important event happens—the steam condenses into water droplets. If the cooling continues down to 32°F (0°C), the water freezes into ice. These structural transformations are called phase changes.

The history of the universe is like the story of the steam, except with six phase changes instead of two. The first four of these happened before the universe was a second old. We'll come back to them after we've learned a little more about what matter is made of. For the moment, let's take up the story of the expanding universe at an age of about one minute.

At this time the universe is a swarm of elementary particles (protons, neutrons, and electrons) and photons of light, all moving and colliding at high energy. If a proton and a neutron combine to form a simple atomic nucleus, the next collision it encounters will be violent enough to tear them apart. Not until the universe turns about three minutes old do things cool down enough for protons and neutrons to bond into a stable nucleus that survives subsequent collisions. This is a phase transition.

These first collisions produce simple nuclei, composed of a single proton and a single neutron. In subsequent collisions, nuclei with more protons form—two protons merge to create the helium atom and just a few three-proton nuclei form the lithium atom. But after about 45 seconds of this, a new effect emerges to shut down the production of nuclei. The Hubble expansion carries the particles far enough apart so that no more collisions occur and no more nuclei are made.

Particles formed in the Big Bang include protons (orange), neutrons (yellow), and electrons (blue). When the universe was cool enough, particles combined and formed atoms (lower right).

So here's one big part of how the universe got the way it is. The universe births itself with nuclei of hydrogen, helium, and trace amounts of lithium. Heavier elements, including the carbon in your tissues and the iron in your blood, are forged later in the guts of stars.

THE ATOMIC UNIVERSE

When do atoms become players in this drama of the universe becoming what it is?

The several-minute-old universe we just left is a hot, expanding gas of nuclei rattling around with free electrons in a soup of electromagnetic radiation. This fourth state of matter is called plasma, which forms whenever electrons are torn from atoms in

THE SIX PHASE CHANGES

Nearly all matter on Earth exists in three states: solid, liquid, and gas. Add or subtract enough energy to a material, and it will undergo a phase change from one state to another. Four transitions are familiar. The remaining two, not so much:

- **Melting:** Solid to liquid
- **Freezing:** Liquid to solid
- **Vaporization:** Liquid to gas
- **Condensation:** Gas to liquid
- **Sublimation:** Solid to gas
- **Deposition:** Gas to solid

Sublimation occurs when the air pressure above a solid is too low to sustain the liquid state of that substance. Everyday air pressure works well to sustain liquid water, but it fails for liquid carbon dioxide. So-called dry ice, for example, which is solid carbon dioxide, goes straight to gaseous carbon dioxide at room temperature and ordinary atmospheric pressure.

Deposition happens when a gas loses energy so quickly it skips the condensation stage and forms a solid. A mild example of this is a frosted forest floor on a chilly winter morning, after water vapor in the air crystallizes into frost on the ground vegetation.

Frost on a winter morning: an example of deposition, the phase change from gas to solid

Neil deGrasse Tyson ✔
@neiltyson

Don't know if it's good or bad that a Google search on "Big Bang Theory" lists the sitcom before the origin of the Universe

💬 4   🔁 391   ♡ 86   1:01 PM - Oct 7, 2010

a gas. This typically occurs at high temperatures. Our Sun, for example—which counts as a high-temperature place—is composed entirely of plasma.

Atoms form when these free electrons hook themselves onto a willing nucleus. But not until the universe was 380,000 years old did the temperature drop enough for freshly formed atoms to remain stable, enduring all subsequent collisions.

During this early period, then, before the temperature fell and atoms endured, the universe consisted of particles with net electric charge—negative electrons and positive nuclei. The radiation that shared the early universe with the plasma interacted violently with the charged particles. If a clump of plasma tried to come together under the influence of gravity, the radiation blew the clumps of matter apart. Think of the radiation as a bunch of cannonballs careening around the plasma, blasting away at these concentrations. And this, of course, meant that galaxies and stars—themselves clumps of matter—couldn't form.

That's why the eventual formation of atoms is so important. Healthy atoms don't have a net electrical charge—all the protons in the nucleus balance the electrons outside it. This property inoculated the atoms against the violent interaction with radiation. The conversion of the charged plasma into a collection of neutral atoms, then, had two effects: Galaxies and stars began to form, and radiation, freed from its condemnation to ricochet among charged particles, escaped. The universe became transparent.

You can experience a similar phenomenon the next time you prepare a glass of iced tea. Watch what happens as you add sugar. At first the iced tea looks cloudy. Sugar in large clumps scatters light, so none of it comes straight through. As the sugar dissolves into electrically neutral molecules, the cloudiness dissipates and the fluid becomes clear and transparent again.

In the same way, once the particles of the plasma became full-blooded atoms, the radiation no longer prevented the accumulation of material under the influence of gravity. The familiar modern universe was being born, and the radiation flew off to become what today we measure as the cosmic microwave background.

BUILDING THE FAMILIAR UNIVERSE

How did the universe transform from an expanding collection of atoms into our cozy world of galaxies, stars, and planets? This question bothered cosmologists for a long time.

People trying to answer this question were inevitably confronted and confounded by an intractable problem. Because radiation decimated charged particles, matter could not begin to clump and form galaxies until neutral atoms showed up and the universe became transparent. By that time, however, the Hubble expansion had spread matter so thin that even the mutual attraction of all the known sources of gravity wouldn't succeed in attracting enough material to make a galaxy. So what happened?

The solution came from a mysterious, unexpected source: something we call dark matter. In the 1930s, the Swiss-American astrophysicist Fritz Zwicky first identified this stuff as part of a cluster of galaxies whose constituent members were moving much faster than their visible sources of gravity could possibly sustain. Thus was born the dark matter problem in the universe. In 1970, American astrophysicist Vera Rubin rediscovered evidence for dark matter during observations of stellar orbits within galaxies themselves.

As the term implies, dark matter does not interact with light, nor does it interact with any other kind of electromagnetic

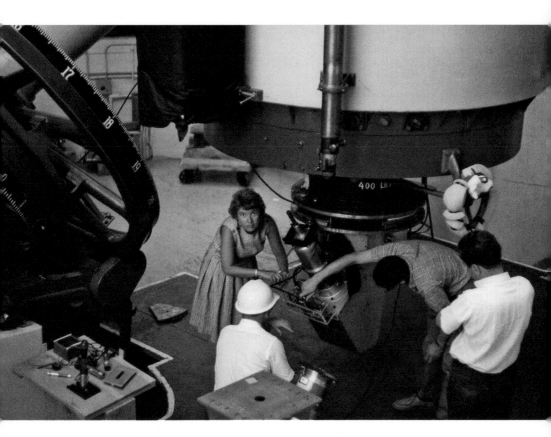

A young Vera Rubin adjusts the radio telescope at Arizona's Lowell Observatory. Her work confirmed the existence of dark matter.

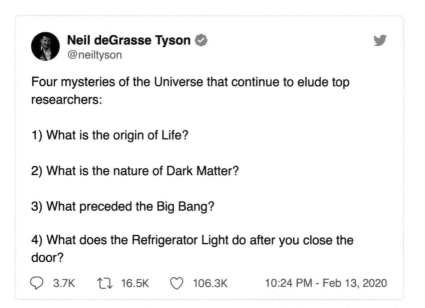

radiation. It does, however, exert a gravitational force. In fact, we have come to understand that dark matter is the source of 85 percent of all gravity observed in the universe—thereby solving the clumping problem.

Recall that matter couldn't clump into galaxies before neutrally charged atoms formed, because radiation in the plasma blows concentrations of charged matter apart as soon as they start to form. Dark matter, however, invisible and impervious to the destructive radiation, accumulated before the universe became transparent. Thus, when atoms formed, they found themselves in wombs of dark matter—within which gravitational attraction could begin and persist.

Imagine you have a bag of marbles, and you begin dumping all those marbles onto a tabletop mottled with deeply drilled holes. The marbles fall effortlessly into these indentations, forming clumps around the holes. The table is the universe. The marbles are ordinary matter, and the holes are the effects of dark energy. All the ordinary matter had to do was fall into the gravitational holes that dark matter had already carved out.

Although we still haven't a clue what dark matter is (the formal term for this state of ignorance is "clueless"), we can measure dark matter's direct gravitational influence on ordinary matter and hypothesize that it primed the early universe for matter to gather and become galaxies, ready to form stars, planets, and people—the universe we know today.

FROM ATOMS TO STARS | Gravity is a beast. Never ceasing, ever enduring, pulling everything together. Stars feel it constantly. At each stage in a star's life, a new strategy deploys that staves off the relentless pull of gravity. The first step in this journey begins with thermonuclear fusion—a foundational process that helped shape the universe as we know it today.

When we last saw our universe, ordinary matter had collected into galaxy-size clouds in gravitational nesting grounds created by dark matter. Within those clouds, matter concentrated more in some places than in others. In other words, galaxies can get lumpy. The extra gravitational force exerted within each lump attracts in the surrounding matter, enlarging the lump ever further. As the lump grows, its gravitational force pulls in even more matter, strengthening the gravitational force further still—and on and on. The simple effect of gravity, then, turns gigantic clouds of scattered atoms and molecules into collections of smaller, more compact objects that eventually become stars and planets.

As gravity does its thing, the gaseous matter collapses and heats up in response. Hot, fast-moving atoms now collide with enough energy to tear off their electrons, turning the gas of the protostar back into a plasma. As the contraction continues, the

600,000,000 TONS The Sun converts this much hydrogen into helium every second.

STELLAR LIFETIMES

You might think that massive stars, because they have more hydrogen fuel, would last longer than smaller stars. Actually, the opposite is true. Because large stars have to work harder to counteract the greater gravitational force, they consume their hydrogen reserve much faster than smaller stars. A high-mass star may live only tens of millions of years, whereas the lowest-mass stars will keep shining for trillions of years.

temperature at the center keeps rising, reaching millions of degrees. Now something new happens.

Protons, all positively charged, normally repel one another by the electric force. But at these high temperatures, they move fast enough to overcome this repulsion and fuse to become larger nuclei. Behold thermonuclear fusion, which generates stupendous quantities of energy.

The mass of the larger nuclei is slightly less than the sum of the smaller nuclei that compose them. That mass gap, as prescribed by Einstein's famous equation $E = mc^2$, liberates the energy that sustains the star. That energy percolates outward with a pressure that, when combined with the pressure within the hot plasma itself, balances the force of gravity and stops the contraction. When this first wave of energy reaches the surface . . . a star is born. Modern cosmology suggests that the first stars began to shine when the universe was about 300 million years old.

Eventually, when a star's nuclear fuel runs out, other lines of defense are employed. Depending on its total mass, the star can end its life as a white dwarf (a cooling ember about the size of Earth) or a dense neutron star (the end product of a supernova

Mystic Mountain, here in an image thanks to the Hubble telescope, is a region of intense winds, churning gases, and energetic star formation.

explosion) a mere 10 miles (17 km) across. In a supernova, atoms created by fusion return to space, ready, willing, and able to seed future stellar systems. The most massive of stars end up as black holes—the ultimate triumph of gravity.

THE NEBULAR HYPOTHESIS

Pierre-Simon, marquis de Laplace was a mathematician—French, of course—whose name is familiar to every scientist and engineer. He rose to fame in France, where Napoleon, a close connection, appointed him as minister of the interior. That appointment lasted only six weeks, by which time Napoleon realized Laplace was a worse than average administrator and sent him back to the academy.

But the French government's loss was astronomy's gain, because Laplace figured out that solar systems like ours arise from the gravitational collapse of a large cloud of interstellar gas and dust. Such clouds are called nebulae (Latin for "clouds"), so the idea became known as the nebular hypothesis.

As already noted, this collapse raises the temperature of the cloud at its center until thermonuclear fusion occurs—but there's another important phenomenon involved. So far, parts of the cloud that had no sideways motion fell straight to the center. All other parts slide into orbit as the cloud continues to contract. The cloud's rotation speed is greatly amplified, just as ice-skaters spin faster when they pull in their arms.

AN EXCHANGE THAT MAY NOT HAVE TAKEN PLACE

When Laplace presented his tome on celestial mechanics to Napoleon, the emperor is supposed to have said, "You have written this huge book on the system of the world without once mentioning the author of the universe." Laplace replied, "Sire, I had no need of that hypothesis."

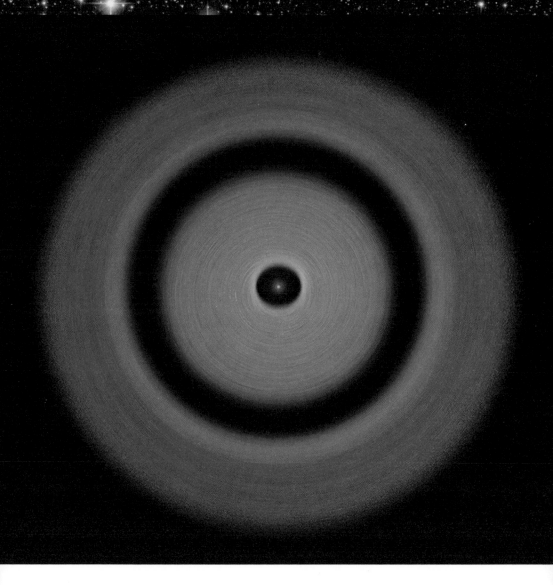

A graphic representation of dust and gas swirling around the red dwarf star TW Hydrae, 176 light-years away. The dark ring suggests a protoplanet forming, gathering up matter as it orbits its star.

The rapid rotation sweeps all remaining debris into a spinning, flattened disk around the newly created star. Within this disk, Laplace suggested, planets would eventually form—all in the same flattened plane, orbiting their star all in the same direction: an elegant model for how the solar system came to be what we know it to be today.

The nebular hypothesis accounts for many features of our solar system and implies that the formation of planetary systems should be common—which suggests, in turn, that Earth may not be the only planet in the galaxy that supports life. Today's spacecraft and telescopes have detected protoplanetary disks around many other newly forming stars. And we now know that planets are so common in the Milky Way that they likely out-number the hundreds of billions of stars themselves.

THE FROST LINE

Once the Sun turned on fusion reactions and started to shine, the rest of the solar system began to take shape. But when we look around the neighborhood today, we see a major difference between the planets. Close to the Sun, we find the rocky worlds—Mercury, Venus, Earth, and Mars, the so-called terrestrial plan-ets. Farther out we find the gas giants, Jupiter and Saturn, and the ice giants, Uranus and Neptune—a family of four collectively called the Jovian planets.

How did two distinct categories of planets come to dominate our solar system? The nebula that eventually became our solar system was composed of two kinds of materials: volatiles and non-volatiles. Volatiles are elements like nitrogen and molecules of water that easily vaporize when heated. Nonvolatiles are materials like sand grains that remain solid at those same temperatures.

While the Sun cranks up its fusion reactions, two things hap-pen. First, the Sun starts to radiate energy into space, raising the temperature of materials close to it. Second, the solar wind—particles cast into space from the Sun's surface—intensifies dramatically. The heat turns any volatile materials into gas, and the solar wind sweeps these gases from the inner solar system. So the only materials left for making planets are things like mineral grains that have not vaporized.

Planets built from this winnowed selection of material are small and rocky. Those farther away from the Sun, in the outer solar system, are large and composed primarily of volatiles. The dividing line between nonvolatile rocky planets and volatile Jovian planets occurs at the "frost line"—a distance from the Sun where the temperature has dropped too low to vaporize volatile materials in the planetary disk.

So simple physics explains some of the salient features of star systems. Or does it? As we shall soon see, the subsequent stages in the evolution of the universe quickly become both complex and curious.

COSMIC BILLIARDS

We once presumed planet formation was simple, and that the solar system itself was surely representative of star systems across the galaxy. We figured that inside the frost line, mineral grains and other solids would stick together to form house-size objects called planetesimals that, in turn, would combine under the influence of gravity to form Mars-size protoplanets. These would then sweep up the remaining debris in the disk and grow to their present size. Farther out, we figured, the Jovian planets must have formed in ways similar to how the Sun formed from an interstellar cloud.

Unfortunately, that simplicity was not to last. By the early 21st century, we finally figured out how to make detailed and

Neil deGrasse Tyson ✓
@neiltyson

Solar System has always been a kind of shooting gallery - or rather, a cosmic ballet, choreographed by the forces of gravity.

💬 28   🔁 146   ♡ 39   3:31 PM - Nov 7, 2011

As this illustration shows, planetary objects swirled into orbit and collided with one another as our solar system took shape.

accurate computer models of evolving protoplanetary disks. And it was anything but simple.

The inner solar system may have been populated by up to 30 planet-size objects whose gravitational interplay resembles cosmic billiards. Some of those early objects collided with one another, shattering into bits. Some were incorporated into other bodies; others plunged into the Sun. Still others were entirely ejected from the solar system on a journey to becoming the most interesting objects in our story: rogue planets. So the eight planets we know and love are just the lucky survivors, with stable, uncontested orbits.

Although a planet kicked out of the solar system is no longer bound by the Sun's gravity, it cannot escape the gravitational field that pervades the entire galaxy. Like the Sun itself, the planet will orbit the galactic center, just as stars do. If every planetary system that has formed since the birth of the Milky Way contributed a few homeless planets to the mix, we are led to an extraordinary conclusion. Not only are there more planets than stars in the galaxy, but the majority of planets may also not be circling stars at all.

PLANETARY MIGRATION

While cosmic billiards played out in the inner solar system, the Jovian planets were caught up in a game of their own. We once presumed the formation of the solar system was smooth and stately.

We were wrong.

In fact, our current understanding has Jupiter caroming around the solar system, sculpting many of the unusual features we see today. Some have nicknamed the scenario we're about to describe "the grand tack," after a sailing maneuver called tacking, which allows a boat to move into the wind by continually changing direction in a zigzag pattern.

Here it goes: Jupiter formed just outside of the frost line over the course of a few million years. Its sustained interaction with the protoplanetary disk sent it spiraling slowly in toward the Sun. During Jupiter's migration, Saturn grew to roughly its present size and began its own inward migration. Smaller than Jupiter, Saturn moves more quickly, and got close enough to Jupiter to exert a significant gravitational force. The gravitational interaction between the two planets and the protoplanetary disk caused

THE PLIGHT OF PLUTO

In the solar system, Pluto has never not been odd. It's a small planet, yet found where ice giants should be. And it travels on an eccentric orbit, embarrassingly tilted relative to the rest of the planets.

Pluto is, in fact, the first Kuiper belt object discovered—not the end of the beginning, but the beginning of the end. Pluto was considered a planet until 2006, when a controversial decision officially demoted it to a dwarf planet. The International Astronomical Union (IAU) laid out three criteria of planetary status:

- **1** The object orbits around the Sun.
- **2** The object is round.
- **3** The object clears its orbit.

Pluto fulfills the first two criteria for planethood, but fails the third. It is a round object that orbits the Sun, but it's not massive enough to have gravitationally dominated its orbit, instead sharing the space with other small, icy objects with similar orbital features, collectively known as Plutinos.

A few astrophysicists, many planetary scientists, and anyone who learned that Pluto is a planet in elementary school remain critical of the IAU planet definition. It includes criteria for where you might find an object rather than for what the object is—which, among other problems, confounds efforts to establish planethood for rogue planets that wander interstellar space.

Also, Pluto, with its active geology, its thin but complex atmosphere, and a possible subsurface ocean of liquid water, might well turn out to be—like Europa and Enceladus—a home for extraterrestrial life.

Composite image of Pluto taken by New Horizons

both to turn around and start migrating outward. By this time, Uranus and Neptune formed, and the gravitational interaction among outer planets moved them to their present orbits.

Strange as it may seem, this complicated choreography among the outer planets actually explains many properties of the inner ones, too, fine-tuning our understanding of how the universe got to be the way it is. Jupiter plowing through the protoplanetary disk was like a bowling ball going through ten-pins. Some material in the disk was pushed into the Sun, while other material was ejected from the solar system; thus Mars and the asteroid belt are much less massive than you would expect them to be. Any protoplanets that might have grown bigger than Earth would have suffered the same fate; thus Earth is the largest possible rocky planet in our solar system.

Finally, the reshuffling of the four Jovian planets at the end of the planetary maneuvering sent a rain of icy comets and debris in toward the terrestrial planets, creating what is called the period of late heavy bombardment. Some evidence shows this may be the source of all water in Earth's oceans.

THE OUTER REACHES

So there's the current story of our well-known eight-planet solar system. What's beyond, and how did it get to be that way?

On New Year's Day 2019, the New Horizons spacecraft flew by an object in the Kuiper belt, the realm of small icy bodies in our solar system far outside the orbits of the familiar planets. The object is formally designated 2014 MU69, with the name Arrokoth, the Powhatan word for "sky." It's one of millions of such objects orbiting the Sun in a puffy disk that extends out beyond the orbit of our farthest planet, Neptune.

Most Kuiper belt objects (KBOs) are composed of frozen volatiles like water, ammonia, and methane. These materials were too distant to be affected by all the shenanigans that went on near the Sun and so remain a repository of the earliest stages of the solar system, like piles of debris left on a construction site after a building is finished.

NAMING KUIPER BELT OBJECTS

By convention, Kuiper belt objects are named for mythological figures associated with creation—but they often acquire nicknames before the official naming process plays out. Eris, for example, is named for the Greek goddess of discord but was nicknamed Xena, after the fictional warrior princess on a popular TV show. Makemake, discovered around Easter time, was nicknamed Easter Bunny before it was officially named for the chief god in the pantheon of the Easter Islanders.

This artwork depicts a swarm of icy bodies in the Kuiper belt, but packs them more tightly than they would truly appear.

In 2003, a planet-size object was discovered among the rubble in the Kuiper belt. Only slightly smaller than Pluto, this object was eventually named Eris. More than a dozen planetary objects like this have been discovered, and there may be many more. Other KBOs bear names like Haumea and Makemake, often an indication that they were discovered using telescope assets on Mauna Kea, on the Big Island of Hawaii. One particularly interesting hypothesis, based on irregularities in the orbits of some KBOs, is that there may be a body out there, 10 times the size of Earth, tugging on them.

And beyond the Kuiper belt lies a distant, yet-to-be-explored region surrounding the solar system on all sides, known as the Oort cloud, also composed of icy objects.

The cozy, comfortable world of the familiar solar system— planets, moons, and asteroids—is actually a tiny part of the whole. The solar system extends far beyond the limits that, until recently, were thought to bound it.

HOW OLD IS
THE UNIVER

SE?

Astrophysics inspires art in this illustration of the first stars birthed in the early universe.

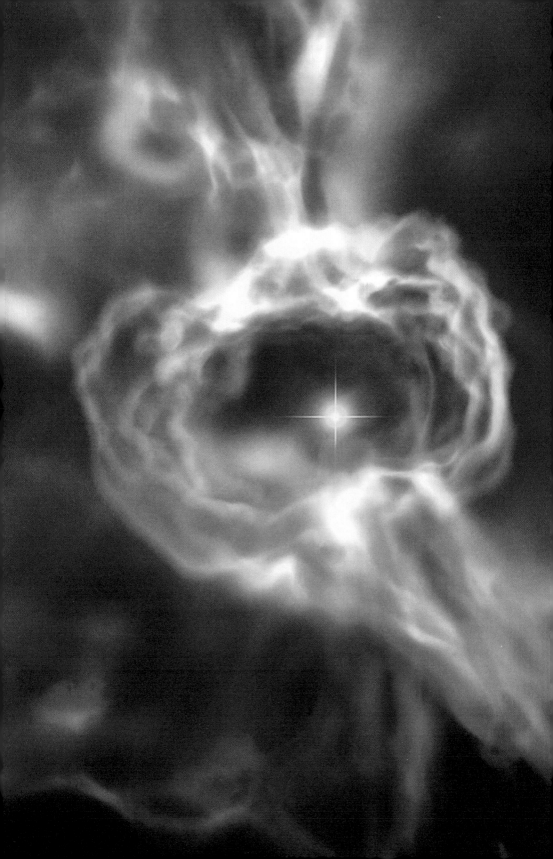

4

Right now, you read these words not as they are now but as they were a couple of nanoseconds ago. That's how long it took light to travel from the page to your eye. In the same way, light from the Sun takes about eight minutes to travel from the solar surface to Earth. (For all you know, the Sun could have exploded five minutes ago, but you'll have to wait another three minutes to find out.)

If we want to know how old the universe is, then we need to detect light that was emitted by the farthest object we can see. The age of the universe is estimated to be either 12.5 or 13.8 billion years old, though most scientists agree on 13.8. (We'll discuss the reason for the difference shortly.) This means that the oldest light we detect today was emitted 13.8 billion years ago. If the universe was static, or stationary, we would say the observable universe is a sphere extending 13.8 billion light-years from the center.

But we don't live in a static universe. We live in one that's expanding. This means that while light emitted from a distant

Earth and space telescope data inform this image of "first light" at the dawn of the cosmos.

galaxy makes its way toward Earth, that galaxy is also moving farther away from us. If we define the observable universe as all the objects we've ever seen or ever can see, then we're talking about a sphere, centered on us, that extends about 45 billion light-years in every direction.

But maybe the universe extends beyond what we can observe—which leads directly to the question: What fraction of the total universe is spanned by the observable universe? If, as some theorists suggest, the observable universe is a small part of the whole thing, then the actual edge, if there is such a thing, remains forever out of reach.

Who knows what lurks beyond our cosmic horizon, let alone how old it is? Even what lurks within it puzzles us. We are still being surprised by what we discover. Each surprise adds information—and more questions.

SURPRISE #1: COSMIC MICROWAVE BACKGROUND

Anything with a temperature above absolute zero radiates electromagnetic waves into its environment. The variety of waves emitted depends on how hot the object is. The Sun, with a surface temperature of about 5000°C, peaks in the visible part of the spectrum. Ordinary objects on Earth's surface, including your body, peak in the infrared—wavelengths longer than visible light. Objects at much lower temperatures peak at even longer wavelengths. The cosmic microwave background hails from when the universe was young.

You experienced this progression of radiative energy the last time you sat around a campfire. While the fire was roaring, the coals at the center may have appeared white hot, emitting light at all visible wavelengths. As the fire died down, the coals turned red—still emitting visible light, but light that had, on average, shifted toward the long wavelength, or red, part of the spectrum.

The next morning the coals no longer glowed, but you could still feel heat if you put a hand over them. The faded but still warm coals were radiating primarily in the infrared.

The universe is like those coals in your campfire. It started out hot and dense and has been expanding and cooling for billions of years. By analyzing the radiation it emits, we can trace its history, and thereby measure the age and also the size of the universe.

In 1964, Arno Penzias and Robert Wilson, two physicists at the Bell Telephone Laboratories in New Jersey, conducted a simultaneously practical yet dull experiment when they accidentally unlocked a secret message from the universe. Satellite communication was new at the time, and the systems used microwaves to send signals. Penzias and Wilson were using an old microwave receiver to scan the skies and see what kinds of random microwave signals might be out there, because that radiation might well interfere with the satellites themselves.

To their consternation, they found that wherever they pointed the receiver, they detected a faint microwave signal, manifesting

WHAT IS ABSOLUTE ZERO?

The temperature and the age of the universe go hand in hand. But to talk about temperature, we need a baseline. Because there's no such thing as cold, only the absence of heat, you can have as much heat as you can muster—but as you take it away, you reach a cold limit: minus 459.67°F (-237.15°C), or zero degrees Kelvin: absolute zero. This state is achieved when no heat energy remains within a substance; the motion of atoms and fundamental particles is as low as it can be.

Temperature itself is a measure of atomic motion. At room temperature, for example, atoms move at the speed of a jet plane, but over tiny distances. Though absolute zero has yet to be achieved in a laboratory, physicists at MIT got pretty close when they cooled sodium gas to within 450 billionths of a degree.

as a hiss in their earphones. In situations like this, experimenters presume the problem to be a fault of their apparatus, so Penzias and Wilson spent a lot of time troubleshooting. They did find that a group of pigeons had nested in the receiver, depositing—as the physicists delicately put it—a "white dielectric substance" on its surface. But even after they shooed away the pigeons and cleaned up their poo, the microwave hiss persisted.

Eventually Penzias and Wilson made contact with physicists at nearby Princeton University, who deduced that the hiss wasn't an artifact of their detector. Instead, the signal emanates from deep space, which is why they detected the hiss in every direction

Arno Penzias and Robert Wilson view their horn-shaped microwave receiver, which made one of the greatest discoveries of the 20th century—the cosmic microwave background.

they turned their receiver. The physicists pointed out that the cooling of a very hot universe, after billions of years, would by now be giving off microwave radiation. Like the campfire coals, the universe was radiating at a wavelength appropriate to its temperature.

This surprise discovery told us that we can see backward in time to events a few hundred thousand years after the Big Bang. In fact, subsequent satellite measurements confirmed this picture, and the discovery of the so-called cosmic microwave background (hereinafter abbreviated as CMB) provides critical evidence in support of the Big Bang theory.

For their work, Penzias and Wilson received the Nobel Prize in 1978.

SURPRISE #2: THE MESSAGE IN THE MICROWAVES

Think for a moment about where the cosmic microwaves came from. You may recall that about 380,000 years after the Big Bang, the universe had cooled to the point that atoms could survive—and that before this time, the matter in the expanding universe was in the form of a plasma, enshrouding all radiation.

When atoms formed, the universe became transparent, and all electromagnetic waves were liberated. Since then, the expansion of space has stretched out those waves, cooling the universe to a temperature that emits microwaves. The cosmic microwave background, then, is a time conduit to the state of the universe when atoms formed.

A composite map reveals tiny fluctuations in the cosmic microwave background (CMB)—the heat throughout the universe left over from the Big Bang.

And this brings us to a surprising fact. It turns out that no matter which direction we look into the cosmos, the temperature of the CMB is the same to one part in 10,000. The entire universe cannot all have a uniform temperature if the entire universe was not in thermal contact with itself. The temperatures from room to room within your fully temperature-controlled home vary by more than this. I don't care how big your home is, it's much, much smaller than the universe.

This is where temperature begins to tell us about the age, and also the size, of the universe. The CMB traces back to the formation of the universe. The 380,000 years between the Big Bang itself and the formation of atoms would have offered plenty of time for any two parts of the universe to cool at slightly different rates, reaching different temperatures from one another.

How, then, can the entire universe know what temperature it should be? And match that temperature so precisely—everywhere? The fact that it's all the same temperature strongly suggests that the universe spent much less time expanding than supposed. We call this the horizon problem—and we will need the inflationary model of the universe to help explain it.

SURPRISE #3: THE INFLATIONARY UNIVERSE

Elementary particle physics is arguably one of the most important developments in modern science, and it becomes critical to our quest to know the age of the universe.

Once upon a time, our newborn universe was so hot that violent collisions broke matter down into its most fundamental forms. The interactions between these elementary bits of matter marked the first step in its evolution. Thus, to understand the biggest thing we know—the observable universe—we must understand the smallest things we know: elementary particles.

One of the earliest breakthroughs in the study of this kind of cosmology occurred in 1979, when American physicist Alan Guth and others developed what has come to be called the inflationary universe, named at a time when the inflation rate for the U.S. economy exceeded 10 percent.

Inflationary cosmology is still being explored and developed, but selected features worked swimmingly. Applying a modern understanding of particle physics, Guth discovered that when the universe was 10^{-35} seconds old, it froze.

Freezing is a familiar form of phase change. We know that some materials, such as water, expand slightly when going from liquid to solid. This is why water pipes burst if they are not protected during a cold snap. Guth found that the universe could also expand stupendously as it went through this particular inflationary phase transition—an expansion of space itself.

Intriguing. But the hypothesis also tells us that before inflation, the universe was much smaller than you would expect just by playing the film backward from the Hubble expansion. In this

ONE PART IN 10,000

To get some idea of what it means for two measurements to vary by less than one part in 10,000, imagine two yardsticks, each purporting to be accurate. If, when you lay them side by side to compare, one is longer than the other by less than the width of a human hair, then they differ by less than one part in 10,000.

pre-inflation period, the universe was tiny, making it trivial for all its parts to come into thermal equilibrium and achieve the same temperature. Once established, this uniformity of temperature would remain imprinted everywhere across the universe during the subsequent period of rapid expansion.

Not only that, but large fluctuations also flatten out and become small fluctuations during rapid expansion. So the inflationary model not only explains why the large-scale universe is at the same temperature with itself, but also why the temperature fluctuations themselves are so small. Thus solving the horizon problem.

SURPRISE #4: TEMPERATURE DIFFERENCES IN THE CMB

To understand the importance of small yet detectable differences in the CMB, we have to remember what the universe was like before the formation of atoms. All matter was in the form of a plasma, with charged particles like protons and electrons bobbing around a sea of electromagnetic radiation. The moment matter attempted to clump together, the high-energy radiation blew the concentration apart. Each event like this produced a wave in the plasma. Because of their similarity to sound waves

Artistic visualization of dark energy propelling the cosmic inflation of the universe

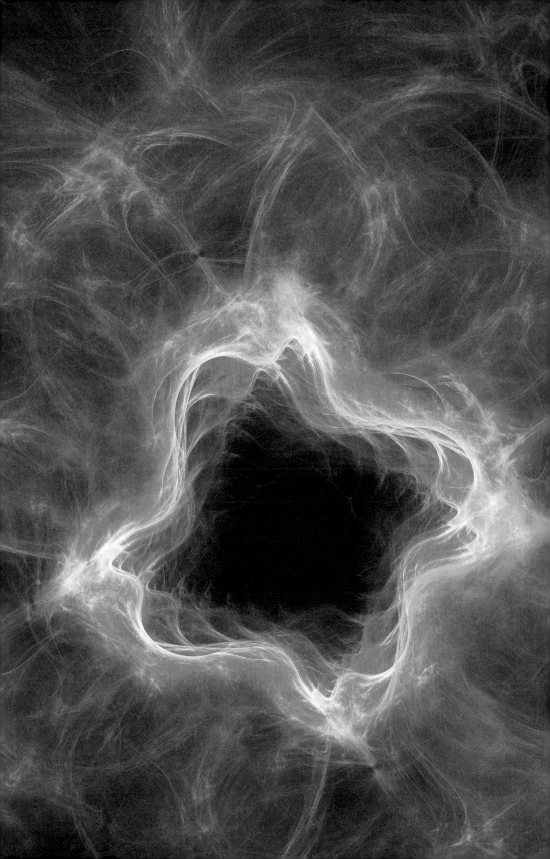

in air, these waves are often called acoustic waves or acoustic oscillations.

The plasma naturally lumped up in some regions more than others. The waves formed a complex pattern in the plasma, with more complexity generated within regions of higher density, such as what you would see if you threw a handful of gravel into a still pond. These patterns propagated through the plasma until atoms formed and the radiation was released.

Meanwhile, the radiation was imprinted with these patterns and thus "froze" into the structure of the universe. The regions of high density, dark matter included, seeded the formation of galaxies. This is why studies of where galaxies are in the universe, together with a detailed temperature map of the CMB, can help us disentangle and decode the history of the universe on our way to finding out how old it is.

Data from the European Space Agency's Planck satellite, which measured temperature variations in the CMB using the

SMALL NUMBERS

The number we have written as 10^{-35} seconds—the age of the universe at this key moment of inflation—is a decimal point followed by 35 zeroes and a 1: 0.00000000000000000000000000000000001. Don't try to picture this time interval—it's smaller than anything in our experience. For reference, the world's fastest computer completes calculations in about 10^{-18} seconds—an eternity compared to the age of the universe when inflation occurred.

best instruments available, combined with data from galaxy surveys, give us a reliable age of the universe: 13.8 billion years.

But don't get too comfortable with that solution. There's another way to measure the age of the universe, and it doesn't agree with that one.

THE ASTRONOMICAL DISTANCE LADDER

The CMB gives us one way to find the age and size of the universe. As often happens in science, however, there are other, completely independent ways of answering the same question. These alternate techniques provide a valuable check on our assumptions and our measurements, because they depend on different properties of the universe. In an ideal world, of course, all methods will yield the same result for every measurement.

One alternative way to measure the age of the universe depends on its large-scale structure. Unfortunately, this technique runs into one of the most persistent challenges astrophysicists face: finding the distance to astronomical objects. Think of the last time you looked at the stars. One that appears faint might be close and dim, but it also could be highly luminous, yet far away. Here we must return to our distance ladder.

Suppose you want to measure how far away something is. If we're talking about a table or a room in your house, an ordinary yardstick would probably suffice. On the other hand, if we're talking about the size of your city, you would probably want another instrument—perhaps the odometer in your car.

In the same way, if you wanted to find the distance to a city on another continent, you might want to use satellite data. Depending on the distance you want to measure, you need to use a different scale of rulers—a different satchel of tools—a different rung on the distance ladder. And, ideally, you would find or establish regions where the different measurement strategies overlap.

That's where you make sure that the two different techniques give the same result. We must ensure that different parts of the ladder join correctly.

We have already encountered the first two rungs of this astronomical ladder. The simplest way to calculate distance is by triangulation, or parallax. Measuring the angle of the lines of sight to a distant object, together with some simple geometry, allows us to know how far an object is from us. The Gaia satellite, launched in 2013 by the ESA, with its hundreds of millions of stellar parallax measurements, has extended this first rung of the distance ladder out to about 25,000 light-years, and it included Cepheid variable stars.

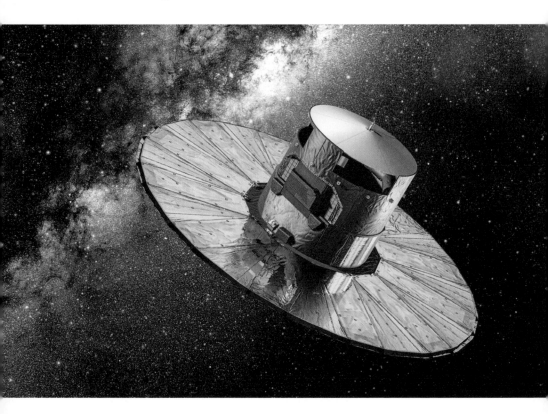

The European Space Agency's Gaia satellite (here in an illustration) orbits the Sun, its mission to create a three-dimensional map of a billion stars.

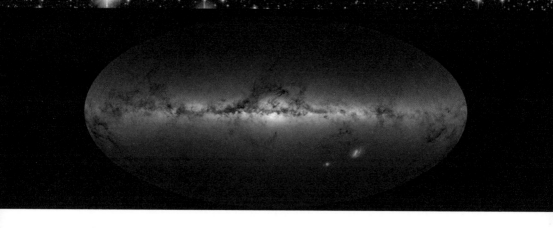

The Milky Way galaxy as viewed by Gaia, capturing nearly two billion stars. Through the middle runs the galactic plane, the flattened disk where most stars reside.

Onto these forms of measurement we graft Henrietta Leavitt's standard candle method, which gets us out to other galaxies. But at a distance out of about 100 million light-years, we are no longer able to distinguish individual stars in distant galaxies, and we must find a new kind of measuring stick—a new standard candle—for measuring the universe's size and ultimately learning its age.

SURPRISE #5: DARK ENERGY

The best way to extend the distance ladder—to add a third rung—is to find another standard candle, but one that can be seen at greater distances. Fortunately, nature has provided just such an object: It's called a Type Ia (*"one-a"*) supernova.

These supernovae occur only in double star systems, where two stars orbit each other and one of the partners is a white dwarf. White dwarfs, though equal in mass to our Sun, are a million times smaller, and therefore a million times more dense. The white dwarf pulls material from its partner until its mass nears 1.4 times that of the Sun. At this "Chandrasekhar limit"— named after Indian-American astrophysicist Subrahmanyan Chandrasekhar—pressures ignite a series of thermonuclear reactions that quite literally blow the white dwarf apart.

For a few weeks, the supernova can shine more brightly than the entire galaxy in which it's embedded. And, since all Type Ia supernovae are exploding white dwarfs of approximately equal mass, their brightness profiles are all the same, fortuitously becoming a standard candle that is visible to great distances.

Starting in the 1990s, astrophysicists began using Type Ia supernova standard candles to investigate the history of the Hubble expansion. Everyone expected the expanding universe

An illustration depicts the double star system RS Ophiuchi, a white dwarf and a red giant in the constellation Ophiuchus. The white dwarf pulls material from its red giant companion as they orbit together, likely to eventually ignite as a supernova.

to be slowing, because of the collective gravitational force of all its constituents in galaxies.

As these things sometimes go, just the opposite was seen. Distant galaxies were dimmer (in other words, farther away) than expected, which means the expansion of the universe was speeding up. An unexpected pressure filling all of space must be pushing the galaxies apart, against their wishes. American astrophysicist Michael Turner coined the placeholder term "dark energy" to describe whatever it is that's causing it. By the way, dark matter pulls matter together, whereas dark energy pushes it apart. So, despite their similarity in names, we're given no reason to think that dark matter and dark energy are at all related.

As we shall see in the next section, a full understanding of CMB data requires the existence of dark energy, a linchpin in the fate of the universe. Even so, we have no idea what dark energy is. Astrophysicists Adam Riess, Saul Perlmutter, and Brian Schmidt, who conducted the deep-universe supernova studies that spawned this discovery, received the 2011 Nobel Prize in Physics for their work.

But wait, there's more. Using this new rung in the distance ladder, a team led by Adam Riess, after pushing the Hubble Space Telescope to the limits of its capabilities, uncovered our next big surprise. They reported a current expansion rate that corresponds to a 12.5-billion-year age of the universe—younger than the 13.8 billion years obtained from CMB data. The difference doesn't seem like much. What's a billion years between friends? Problem

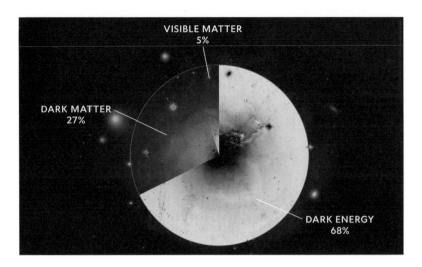

Matter as we know it composes a mere 5 percent of the universe.
We can measure the rest, but remain clueless about what it is.

is, both measurements have high accuracy and excellent precision granted by their own methods and tools. Their margins of error do not overlap, but they both cannot be right. So, either one is wrong and the other is right. Or both are wrong.

Now what?

THE TENSION

We now confront two different ways to estimate the age of the universe. Against all expectations, the two methods report different results—13.8 billion years from the CMB analysis, 12.5 billion years from the distance ladder extension.

The tension between these two competing solutions can soften upon citing a simple fact: There is no such thing as a perfect measurement. No matter how precisely something is measured, there is always a limit to how accurately a result can be known.

Suppose you wanted to measure the size of the room you are in right now. You would probably use a ruler or a tape measure to do the job. No matter how careful you are, however, there is a

fundamental limit to how accurately you can make the measurement. An American ruler illustrates this well. On it, there will be a smallest interval—typically 1/16". If you use a ruler divided this way, there is no way to tell whether the room is 10' 5/32" or 10' 6/32" long. And if your ruler were graded in 1/32" intervals, you would not know if your room were 10' 11/64" or 10' 12/64" long. There is, in other words, a fundamental limit on how accurately you can know the dimensions of your room—and therefore a quantifiable uncertainty in your experiment.

What about measuring the size of a continent? The continent ends at the shore, but what is the shore? The waterline ebbs and flows. Even the exact measurement of low tide is

THE RIGHT QUESTION

If two highly reliable measurements of the same phenomenon give two different answers, and if that difference appears to be reproducibly intractable—even after all assumptions were tested and retested, and after all experiments were verified and reverified by scientific allies as well as competitors—then there's a chance the question that spawned the measurement might not have the meaning you think it should. What is the temperature of love? Where is the edge of Earth's surface? What kind of cheese is the Moon made of? All of these questions are legitimate English-language sentences, with nouns and verbs in the right places, but they have no physical meaning, because the very questions themselves are flawed. Could "How old is the universe?" be one of those questions and we don't yet know it?

never replicated. Because of the Moon's varying distance from Earth, and Earth's varying distance from the Sun, some low tides are lower than others.

At some point, everyone must agree on a number that is close enough.

No matter how accurate your apparatus, no matter how carefully you observe, your measuring equipment will always involve a smallest interval—and, consequently, an uncertainty in your last decimal place. In science, this uncertainty represents the range of values the experiment might produce, if performed multiple times. When you measured your room, you might report a length plus or minus $1/16$ of an inch to account for that uncertainty on your ruler.

The size of the measuring scale is only one of many sources of uncertainty in real experiments and observations. Sometimes there is an undetected flaw in the equipment that has nothing to do with a measurement stick. For example, in 2011, physicists at CERN in Geneva, Switzerland, announced the detection of a particle that traveled faster than light. Had it been true, this result would require a major reworking of Einstein's theory of relativity. But in the end, a badly connected fiber-optic cable was to blame. In this case, the limit on accuracy was set by a problem in the apparatus, and not by any uncertainties in measurement. This is not uncommon, and explains why Penzias and Wilson spent so much time troubleshooting their apparatus before announcing the discovery of the CMB.

Sometimes, the uncertainties are purely statistical. If you want to estimate the average height of your country's population and measure only 10 people on the day a basketball team comes

The ATLAS detector at the CERN Large Hadron Collider forces subatomic particles to collide at high speeds, creating a spray of debris that may contain undiscovered particles.

through town, you can expect results of low accuracy. Pollsters know this. That's why they try to reach a thousand people and not just 10. Even then, their margin of error is typically +/- 3 percent. Measure a million people from the same population, and both accuracy and precision improve.

Uncertainty can also arise because of mistakes in the analysis of measurements. If you thought your ruler, purchased in the United States, was measuring in yards when it was actually measuring in meters, which are 9 percent longer, your answer would be off by 9 percent, regardless of how precisely you measured. Similarly, an unknown background effect—if the room you measured didn't have straight walls, for example, and you didn't know it—can mean your results will be wrong, no matter what.

No doubt about it. Conducting good science is messy business.

When the two different ages of the universe obtained by the two different methods first reared their heads several years ago, scientists paid little mind. At that time both measurements had large uncertainties. And, more to the point, those uncertainties overlapped—meaning the universe could have an age somewhere in that overlapping range. As time went on, however, both groups improved (that is, reduced) their uncertainties, and the difference could no longer be ignored. The CMB age is now quoted as 13.799 +/- 0.021 billion years, while the supernova age is 12.5 +/- 0.3 billion years. If the tiny uncertainties offer no place

Neil deGrasse Tyson ✔
@neiltyson

Unlike the Aether -- hypothesized but never detected -- DarkEnergy is a measured entity. We just remain clueless what it is.

💬 57   🔁 138   ♡ 34   3:43 PM - Oct 5, 2011

ACCURACY VERSUS PRECISION

A digital clock that displays time to 1/100 of a second is precise. In fact, it's more precise than most people will ever require in their lives. But suppose, just suppose, the clock was six minutes fast and you did not know it. That would be a highly precise but wholly inaccurate clock. In science we first seek accuracy: Is the answer right at all? Is it in the ballpark? Next we try to improve on that accuracy by making the measurements more and more precise.

to meet in the middle, then this discrepancy is likely due to some unknown effect in the analysis of the various measurements. In the words of Adam Riess: "This mismatch has been growing and has now reached a point that is really impossible to dismiss as a fluke."

SURPRISE #6: DARK MATTER

Up to this point in our quest for the size and age of the universe, we have been talking mostly about familiar forms of matter, such as the protons and other particles found in your everyday atoms. You can, in fact, think of the entire history of science as an attempt to understand the properties of ordinary matter. But we would ultimately learn more was hiding in the universe than meets the eye.

Beginning in the 1930s, astronomers found clusters of galaxies that seemed to be held together more strongly than could be explained by the gravitational attraction of their visible matter. And in the 1970s, American astrophysicist Vera Rubin found the same phenomenon within galaxies themselves. Stars in orbit around a galactic center were moving much faster than they ought to, given the gravity of all the stars present and accounted for. The only way to explain this result was to assume that the

Neil deGrasse Tyson ✔
@neiltyson

Add up all we know about matter & energy and it accounts for less than 5% of what drives the Universe.

What we call Dark Matter & Dark Energy comprise the rest, yet we know nothing of them, other than they exist, leaving the astrophysicist delightfully befuddled, for now.

890   2.7K   21.8K   4:58 PM May 17, 2020

entire visible galaxy—the part made of familiar matter—was encased in a huge sphere of mysterious material that exerted a gravitational force but did not radiate or interact with electromagnetic waves. This material was dubbed "dark matter."

In the half century since Rubin's measurements, evidence for dark matter has been verified in many different cosmic environments, and it turns out to be crucial to our understanding of how galaxies form, itself vital to understanding the age and size of the universe. We can, in fact, characterize the state of our knowledge of dark matter as follows:

- **We know it exists.**
- **We have no idea what it is.**
- **Perhaps it should instead be called "dark gravity."**

FORMATIONS OF THE UNIVERSE

It's difficult to see the universe in three dimensions—which is, of course, essential to knowing its size. Our first attempt to develop a comprehensive 3D picture of galaxies across space was through redshift surveys. We already know where a galaxy resides on the 2D sky. But after we know a galaxy's redshift—the velocity at which it moves away from us—we then employ

Hubble's law to calculate the distance to the galaxy, completing our knowledge of its cosmic location in three dimensions.

Modern surveys can take advantage of advanced electronics and measure redshifts of hundreds of galaxies at a time. Consequently, modern redshift surveys are much more comprehensive than their predecessors. In 1982, Harvard's Center for Astrophysics used redshift survey techniques and cataloged 2,200 galaxies. By 2007, the Sloan Digital Sky Survey, thanks to advancing technology, published redshifts for more than a million galaxies, offering an ever more comprehensive three-dimensional picture.

When you lay out where all these galaxies are in space, they are not evenly spaced across the cosmos. Instead, a strangely ordered universe reveals itself.

Imagine you have a huge sponge and a big knife. If you use the knife to slice through the sponge, you will see pockets of empty volumes bordered by sponge. In the same way, when we look at redshift surveys, we find empty spaces, sensibly called

THE LUX EXPERIMENT

If the Milky Way is indeed encased in a sphere of dark matter, then the motion of Earth should create a wind of dark matter particles continuously sweeping through and around us. Because dark matter appears to interact only through the relatively weak force of gravity, this wind will interact very seldom with ordinary matter, most of the time just passing through without a trace.

In the Sanford laboratory, located a mile underground in an old South Dakota gold mine, a phone booth–size container of liquid xenon has been outfitted to detect any rare collisions between the dark matter wind and the xenon atoms. This is the centerpiece of the LUX (Large Underground Xenon) experiment. As of this writing, no dark matter particles have been detected in this or any other experiment.

Neil deGrasse Tyson ✔
@neiltyson

Nothing to tweet today, except for all those who wanted more space, the Universe continues to expand at about 70 kilometers per second, per Megaparsec.

💬 969   ⟲ 5.2K   ♡ 43.4K   9:54 AM - Mar 29, 2020

Our solar system (lower right) is part of the Milky Way galaxy (lower left),
which is part of the Local Group of galaxies (upper left),
which is part of a galactic supercluster (upper right).

THE GREAT WALL

The largest structure detected in the observable universe is a supercluster of galaxies called the BOSS Great Wall, named for the study that spotted it, the Baryon Oscillation Spectroscopic Survey. This enormous cosmic webbing spans a billion light-years and looks like a grand honeycomb.

voids, surrounded by threadlike filaments and sheets whose structural elements are galaxies.

Welcome to the large-scale structure of the universe.

To illustrate our place in this grand assemblage—and to answer the intertwining questions "How big is the universe?" and "How old is the universe?" along the way—consider this: Earth is part of the solar system, and the Sun is part of the Milky Way galaxy, a structure some 100,000 light-years across. The Milky Way, in turn, is part of the Local Group of galaxies, a structure some two million light-years across. The Local Group is part of a Virgo supercluster of galaxies some 750 million light-years across. Finally, this supercluster is part of the web of structures surrounding the voids previously described.

A long-earned but nonetheless powerful indication that the shape and form and contents of the universe are knowable—not only in the here and now, but also in the there and then.

WHAT IS THE MADE OF?

UNIVERSE

The quest to uncover answers to one of the most fundamental questions in science can begin by simply looking around you. A superficial peek gives the impression of a universe made from a zillion different materials, each different and distinct, obeying a bazillion different rules.

Suppose someone asks you to define the basic building block from which libraries are made. From the outside, it's clearly made of bricks or some other sturdy construction material. But if you manage to open the front doors and go inside, you will see shelf after shelf of books, so your second response might be that books are the basic building blocks of libraries.

But a library is not just a random pile of books. Classification schemes tell you how the books are arranged—biography, poetry, fiction, and so on. As you probe deeper, you must continue to modify your answer to the question about the basic building blocks of libraries.

Now pick any book off any shelf. Open it, and you see a collection of words. Nearly all the books, in fact, are made up of

To ask what the universe is made of, we must first look deep into space—
and then deep into the nature of matter.

Neil deGrasse Tyson ✔
@neiltyson

Some of my best friends -- actually all of my best friends -- are made of chemicals.

💬 962   ↻ 3K   ♡ 33.5K   11:14 AM - Mar 9, 2020

words—so we have to change our answer to say that the basic building block is the word. And there is a set of rules called grammar that tells us how to combine words into sentences, sentences into paragraphs, paragraphs into chapters, and all those ultimately into books.

But wait. There's yet another layer to be revealed. Some words and word combinations appear in the books of some libraries and not in others, revealing the existence of languages in the world, and perhaps specialized libraries. And for most languages, we quickly discover that words are made from letters, assembled according to rules of spelling that tell us how to combine letters into words. Further, in our digital age, we could also go down one more layer and represent letters by a string of 0's and 1's, with a corresponding rule that tells us how to sequence the 0's and 1's to make letters.

So our question about the basic structure of libraries leads us down a rabbit hole to a picture far more complex than we might have first anticipated or even imagined.

Our search for the basic structure of the universe does the same.

THE BIRTH OF CHEMISTRY

During the Middle Ages, a group of researchers pursued a field they called alchemy. In the popular imagery, they often look like Merlin protégés or refugees from a *Harry Potter* movie. Indeed, some of the more unscrupulous members of the profession got wealthy by promising their patrons they would find ways to turn base lead into precious gold. False promises notwithstanding, the alchemists contributed to scientific progress, and over the centuries, accumulated scads of qualitative information about chemical reactions.

All this knowledge landed on a scientific footing with Antoine Lavoisier and his wife, Marie-Anne Lavoisier, members of 18th-century French nobility. They introduced precision

Medieval alchemists kept good laboratory notes, characterizing elements in ways that informed the future science of chemistry.

to

measurement into chemistry, and Antoine was the first ^show that the total mass in a chemical reaction does not change, which became the law of conservation of mass.

From our point of view, however, Lavoisier's most important discovery was that such a thing as elements exist at all. There may be countless kinds of materials in the world, but most of them can be broken down into smaller constituents by chemical means. Burn wood or dissolve a metal alloy in acid, and you transform the meta-materials into base ingredients.

Some materials, however, cannot break down in this way. The black carbon that is left when wood burns, for example, can be combined with other materials to make something more complicated, like carbon dioxide, but it cannot be broken down into anything simpler. Carbon, then, is an example of what we call a chemical element. Although chemists at the time knew thousands of different materials, they knew only a few elements. In 1776, for example—a year familiar to Americans—among all cataloged substances, only 22 pure elements were known, 12 of which had been identified by the ancients.

By the end of the 18th century, chemists recognized some astonishing regularities among the chemical elements. One of the most important of these is the law of multiple proportions, which states that the relative weights of different elements in a given material will always be the same, no matter where the material comes from. The weight of oxygen to hydrogen in water,

Neil deGrasse Tyson ✔
@neiltyson

Sometimes I wonder whether the Universe can make something more complex than itself.

💬 2.3K   ↻ 8K   ♡ 47.3K   11:23 PM - May 18, 2018

GENIUS INTERRUPTED

Unfortunately, Antoine Lavoisier did not live to participate further in the science of his generation—a bad time in France to be a member of the nobility and to be a scientist (plus, he was connected to an unpopular tax collection company). Lavoisier was sent to the guillotine in 1794, during the French Revolution's Reign of Terror. Pleas for leniency are said to have come from all over Europe, with no success. Fellow astronomer Joseph-Louis Lagrange lamented, "It took them only an instant to cut off that head, and a hundred years may not produce another like it."

for example, will always be 8 to 1, whether the water comes from a tropical island or a melting glacier.

With this arsenal of discoveries, scientists were poised to go ever deeper into the question of what the universe was made of.

WHERE DID THE ELEMENTS COME FROM?

We know the number of protons in the atomic nucleus of each chemical element. We also know that the three-minute-old universe briefly produced nuclei of atoms with up to three protons: hydrogen with one, helium with two, and lithium with three.

Where did all the other elements come from?

To answer this question, one must return to the formation of the solar system. As gravity compresses the material into what will become the Sun, the temperature in the core reaches millions of degrees, triggering nuclear fusion. After a few intermediate steps, four hydrogen nuclei fuse together to form a helium nucleus, as well as some miscellaneous particles plus energy. The Sun then, like most stars, generates energy by converting hydrogen into helium.

When nearing the end of its life, a star exhausts its supply of hydrogen in the core and contracts, briefly surrendering to the inexorable force of gravity. Temperatures in the condensing core rise further, enough for three helium nuclei (with two protons each) to fuse and form a carbon nucleus (with six protons), along with its intended neutrons. Thus the ashes of one nuclear fire serve as fuel for the next, empowering stars to forge elements heavier than those the young universe could.

A star like the Sun isn't massive enough to fuse nuclei past those of carbon. But via the solar wind, it casts some of its hard-earned elements into space. More massive stars, however, can continue fusing nuclei through carbon all the way up to iron (with 26 protons)—the ultimate nuclear ash, and the end of the line for nuclear fusion reactions that generate energy. The iron accumulates in the core of the star—but when the star tries to fuse iron, the nuclear reactions absorb energy: a bad situation. Stars are not in the business of absorbing energy.

The star rapidly collapses under its own weight, with no source of energy to balance the impending disaster. The collapse spawns a titanic explosion we call a supernova. In the resulting maelstrom, with plenty of surplus energy to get the job done, elements all the way up to uranium (with 92 protons) are produced.

Uranium is the heaviest naturally occurring element in the periodic table. All the rest, up to oganesson (element 118, named for Russian physicist Yuri Oganessian), have been produced only in laboratories. How? If you accelerate a heavy nucleus to high speed and let it collide with a target, the subsequent reshuffling of protons and neutrons might form a few atoms of a new element.

THE ISLAND OF STABILITY

While the heaviest superheavy elements produced in the laboratory typically decay in fractions of a second, nuclear theorists predict that when we finally produce element 126, we will reach a so-called island of stability and thus herald a new class of chemical elements that might form the basis of a new kind of chemistry.

63 The number of chemical elements Mendeleev knew about. Today, it's 118 and counting.

For example, we have only ever produced 75 single atoms of copernicium (element 112, named for Polish astronomer Nicolaus Copernicus) in our laboratories. The attempt to produce still more massive superheavy elements remains a thriving cottage industry in the world of nuclear physics.

THE NEW ATOMIC THEORY

John Dalton was an English schoolteacher whose first love was meteorology, a subject that includes the study of chemical reactions in the atmosphere. In 1808, he introduced the modern atomic theory.

In Greek, the word *atom* means "that which cannot be divided"—and indeed, Dalton thought of atoms as indivisible. He proposed that there was an atom corresponding to each chemical element. All the atoms of a given element are identical, and atoms of different elements are different from one another. In this plan, assortments of atoms combine to form the range and diversity of materials that comprise the natural (and unnatural) world. This explains why most materials can break down—it's just a matter of separating the atoms. Only then, Dalton argued, can you no longer take things apart.

Dalton's model explained many of the features of materials—features that were constant, predictable, and recognized but before that unexplainable. Water, for example, consists of one oxygen atom bound with two hydrogen atoms, no matter where it is found. A single oxygen atom weighs eight times as much as two hydrogen atoms, so the 8-to-1 ratio in weight between the two was a simple feature of how atoms combine.

Neil deGrasse Tyson ✔
@neiltyson

Some elements don't interact with others. NobleGases the Brits called them, tainting the PeriodicTable with their ClassSystem

💬 52   🔁 279   ♡ 57   1:42 PM - Nov 4, 2011

As time went by, scientists continued to discover more chemical elements. A 19th-century Russian chemist, Dmitri Mendeleev, encountered the problem of organizing the known chemical elements while writing a textbook. The solution? Mendeleev created the modern periodic table of the elements, still a standard icon in the chemistry classroom. The elements get heavier as you read from left to right in any row, and they have similar chemical properties if you read down any column. In his original table, he left some blank spaces to make things work. He expected that these spaces would be filled as new elements were discovered—and indeed, subsequent discoveries validated his ideas.

Although scientists knew that the periodic table was a valid arrangement, they had no idea why it worked. Not until the 1920s and the discovery of quantum physics could a deep understanding of the table emerge.

DISASSEMBLING THE ATOM

By the late 19th century, scientists uncovered a world of relative simplicity—a world where a small number of indivisible atoms, shuffled around in different combinations, produced the world that is immediately available to our senses. If complex materials

The actual atom, so different from what John Dalton imagined, is dynamic and elusive, as conveyed in this illustration of the nucleus and its family of electrons.

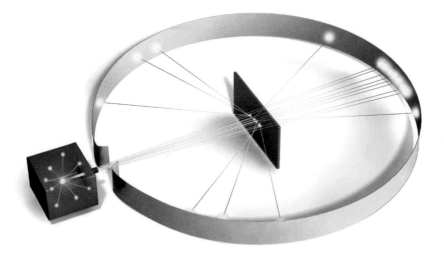

Rutherford's experiment, shown in this diagram, shot particles at gold foil and observed how they scattered. His results showed that most of the mass in the atom is concentrated in the nucleus and that most of the atom itself is empty space.

are the books in our metaphorical library, atoms are the words filling the pages.

This comfortable simplicity began to unravel in 1897, when the English physicist J. J. Thomson performed an experiment that revealed the existence of totally unexpected particles: electrons, negatively charged constituents of matter that shake loose from atoms readily and easily. The problem was that Dalton's atomic theory, which held atoms to be indivisible, had no room for something like this. If an electron can rip apart from an atom, an atom is clearly divisible. For a while, theorists dealt with Thomson's result by picturing the atom as something like a raisin bun, an amorphous, positively charged material embedded with electrons.

In 1911, however, New Zealand physicist Ernest Rutherford announced the results of an experiment that put an end to the idea and crafted our current understanding of the atom. Rutherford aimed a stream of particles—subatomic bullets—toward an extremely thin sheet of gold and observed how they scattered.

Gold is the element of choice in this experience for how thin it can be hammered. In this case, just a few thousand atoms thin. If the atom were really like a raisin bun, this would be like a bullet fired at a cloud and effortlessly passing through.

What Rutherford found was that only about a 10th of one percent of his subatomic bullets bounced back at him. If you shot a round of bullets through a cloud, and one of them came ricocheting back toward you, what might you assume? Something stronger than your bullet was lurking within that cloud.

Similarly, the only explanation for Rutherford's unexpected result was that all of the mass of the atom was concentrated in a small structure at its center—a structure Rutherford called the nucleus—with the electrons in orbit around it. The particles that bounced back had hit the nucleus, while the others had simply passed through the mostly empty cloud of orbiting electrons.

Rutherford called the nucleus of the hydrogen atom the proton, meaning "first one." But he knew that heavier atoms had to have more than protons in their nuclei. The nucleus of the oxygen atom, for example, has eight protons in its nucleus but weighs 16 times as much as hydrogen. Rutherford predicted we would find another particle that had about the same mass as the proton but no electric charge. He called this hypothetical particle the neutron, or "neutral one," and that particle was indeed discovered, in 1932 by British physicist James Chadwick.

AN UNCONVENTIONAL SCIENTIST

Ernest Rutherford is the only scientist we know of who did his most important work *after* he received the Nobel Prize. He got the prize in chemistry in 1908 for identifying some of the products of radioactive decay—an important advance, but not in the same category as discovering the structure of the atom, which happened later.

Once again, we find complexity reduced to simplicity. The universe was built from three particles: protons and neutrons, which make up the nuclei of atoms, and electrons in orbit around them. Atoms then combine to form all the materials we see.

But this simplicity would not last. Even deeper layers of matter lurked just out of view.

WHO ORDERED THAT?

The agents that destroyed the simple proton-neutron-electron picture of the universe came, literally, out of the sky. Earth is constantly bombarded by streams of particles called cosmic rays—mostly protons, mostly from the Sun. These particles have energy high enough to penetrate and tear apart the nucleus of atoms in Earth's atmosphere. Examining the debris of such collisions opened a hidden door to the world inside the nucleus.

This technique of probing the nucleus by examining the debris of particle collisions has been the main tool available to physicists since the early 20th century. The American physicist Richard Feynman likened the method to dropping a Swiss watch off the Empire State Building and then trying to figure out how the watch worked by looking at the pieces on the sidewalk. Sounds messy and clumsy, but you had no other way of looking inside the watch.

By the 1930s, physicists devised a method to harness cosmic rays for this purpose. Physicists deployed detectors at laboratories on mountaintops, such as Pikes Peak in Colorado, where they intercepted the incoming cosmic rays to track their interactions. And that's when the confusion began.

Unexpected things started showing up in the cosmic ray experiments. First there was a particle with the same mass as the electron but a positive electrical charge—the first example of antimatter, aptly named the positron. Then there was a particle like the electron but 200 times heavier and living much longer than it should have. With a half-life of only 1.5 microseconds, it should not have been able to reach Earth's surface. This unexpected particle, named the muon, famously prompted American physicist I. I. Rabi to voice the plaintive cry, "Who ordered that?"

The list grew. Protons have nearly two thousand times the mass of electrons, yet researchers found brand-new particles with a mass between protons and electrons and called them mesons, after the Greek *mesos* for "middle." They found particles even heavier than protons and called them hyperons.

After physicists built particle accelerators with controllable collisions, they no longer needed to rely on random pulses of gamma rays from space. And the list of particles grew even longer. Most of the new particles existed so briefly that they could barely get from one side of the atomic nucleus to the other before disintegrating. Clearly, the nucleus is not the passive bag of marbles labeled "proton" and "neutron" that scientists once thought, but rather a seething cauldron of myriad short-lived particles.

Physicists had opened a whole new box of mysteries. Yet another layer of organization in the universe.

THE COMING OF THE ACCELERATORS

To study the interior of the nucleus, we need very high-speed tools to smash them and very high-precision means to analyze the train wreck. Cosmic rays provided these probes to us for free. Their great disadvantage, however, is our inability to control either the energy or the time of arrival. Particle accelerators granted us the freedom to manufacture and control our own probes.

The history of accelerators begins in the 1930s with Ernest Lawrence at University of California, Berkeley, where he developed one of the first particle accelerators: the cyclotron.

Particle accelerators like cyclotrons depend on a key property of elementary particles. Inject a moving electrically charged particle into a magnetic field, and the path of the particle curves around, eventually forming a circle. Lawrence realized that if you start with a magnet, slice it through the middle, and separate the two pieces by a small gap, he could accelerate positively charged protons as they passed through the gap, propelled by the force of the magnets, thus forming a proton beam that could be fired at a target. Lawrence's first cyclotron would fit in the palm of your hand. By the 1950s, though, the Berkeley cyclotrons had grown to the size of a sports arena.

The next step in accelerators was the synchrotron. Instead of a single massive magnet, these machines configure a series of long curved magnets in a ring. The circle can be yards or even miles in circumference. The particles accelerate along an evacuated path, and then the strength of the magnets can be actively adjusted to keep the faster-moving particles in the vacuum chamber.

The Large Hadron Collider (LHC), the largest accelerator in the world today, is located near Geneva, Switzerland. The word "hadron" refers to any of the particles you might find in the nucleus of the atom. With a vacuum chamber about 17 miles (27 km) around, the accelerator is buried underground, to shield those living above the tunnels from any harmful radiation.

The Large Hadron Collider is not one, but two synchrotrons, with one proton beam traveling clockwise, the other traveling counterclockwise. The two beams can collide at specific points to exploit the double kinetic energy available in head-on collisions. In the

Lawrence's first cyclotron, circa 1931, less than five inches across

Neil deGrasse Tyson
@neiltyson

Top 4 collaborations of Nations: 1) The Waging of War, 2)
International Space Station, 3) Large Hadron Collider, 4)
Olympics.

💬 124   🔁 1.5K   ♡ 421   8:16 PM - Jul 27, 2012

Ernest Lawrence and Stanley Livingston, circa 1946, adjust a newly designed
synchrocyclotron, so large that a building had to be constructed
on the Berkeley campus to house it.

$25 The cost of the first cyclotron Lawrence built at Berkeley
(about $450 today)

A computer simulation portrays an experiment performed by the
ATLAS detector of the Large Hadron Collider. Two beams of subatomic
particles accelerate and collide, creating a shower of particles, some new.
A human figure has been added to convey scale.

$4,750,000,000

The cost to build the Large Hadron Collider

spray of particles created by these collisions, physicists uncloak that next level in the eternal quest to know the structure of matter in the universe.

THE COMING OF THE QUARKS

When scientists introduce a new concept, they choose between two naming strategies. They can take an existing word and give it a new meaning—for example, the word "work," which has a precise meaning in physics not reflected in everyday usage. Alternatively, scientists can invent a new word completely—which is just what physicists did when they opened the next door.

By the late 1960s, elementary particles had proliferated. One group of physicists at Berkeley maintained a register of reported particles and periodically published a summary tome describing the hundreds of known particles. So jaded did some members of the physics community get that an editorial in a prominent physics journal declared that if all you've got to report is a new particle, don't send the manuscript to us—they weren't interested. Even Enrico Fermi, the creator of the world's first nuclear reactor, said, "If I could remember the names of these particles, I would have become a botanist." The physicist's search for simplicity had led to a world of complexity.

It came as a great relief, then, when the American physicists Murray Gell-Mann and George Zweig showed that the proliferation of particles could be explained simply if we went one more layer down. They showed that all the elementary particles could be thought of as combinations of three particles that were more elementary still. Gell-Mann named these particles "quarks," a tip of the hat to a line of wordplay in James Joyce's *Finnegans Wake*: "Three quarks for Muster Mark."

The idea was that different combinations of these quarks correspond to different particles in the particle zoo, just as different combinations of atoms produced different materials in Dalton's atomic theory. These three quarks are whimsically called "up," "down," and "strange." Unlike the elementary particles, however, they have fractional electrical charges—two-thirds or one-third the charge of the electron or proton.

Subsequent experiments have shown that there are actually six quarks in nature, though the original word remains (and *Finnegans Wake* has not been revised). In the beginning, scientists searched strenuously to find free quarks in nature or to produce them at accelerators. When these searches failed, theorists realized that once quarks lock into a particle, they're stuck for good—a property called "quark confinement."

Here's how it works: Try to pull two quarks apart and, like stretching a rubber band, doing so feeds energy to the system. Eventually you succeed in breaking their bond, but the energy you invested to accomplish this task is exactly the energy necessary for $E = mc^2$ to create two more quarks, one for each freshly broken bond.

Today the six-quark model is fundamental to our ideas about the structure of matter, and marks the deepest layer we have yet

---

WHAT NEXT?

The next big project for particle accelerators is the International Linear Collider (ILC). As the name implies, this machine will accelerate beams of electrons and positrons down long straight tubes, allowing them to collide head-on when they reach maximum energy. The machine will be long—about the distance you would run in a marathon. No cost estimates have been made at this point.

---

to reveal on the operations of nature. But the inevitable question remains: Is there yet another layer?

GLOSSARY OF PARTICLE PHYSICS

Once we get into the weeds with elementary particles, we start running into a lot of strange and whimsical words. We assure you, these are all perfectly scientific:

■ **Hadron** | Any of the hundreds of particles that can exist inside the atomic nucleus. The word means "strongly interacting ones." The most familiar hadrons are the proton and neutron.

■ **Baryon** | Hadrons that, like the proton and neutron, are made from three quarks. The word means "heavy ones."

■ **Meson** | Hadrons that are made from a quark–antiquark pair. The name means "intermediate one." The first mesons discovered were intermediate in mass between protons and neutrons.

■ **Lepton** | An elementary particle not normally found in the nucleus. The name means "weakly interacting one." The most familiar lepton is the electron. There are six leptons—the electron and two other more massive particles like it and three types of neutrinos.

■ **Neutrino** | A lepton that has no electric charge and almost no mass. The name means "little neutral one." Neutrinos are copiously produced in nuclear reactions in stars, and detecting them has been a major activity in modern astrophysics. There are three different types of

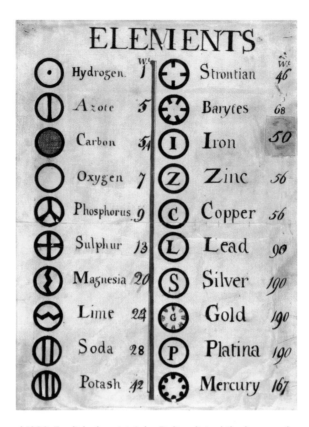

Around 1800, English chemist John Dalton listed the known elements
according to his estimates of their weights.

neutrinos, corresponding to the three masses of leptons
described previously.

■ **Up and down quarks** | The quarks that make up protons
and neutrons.

■ **Quark color** | A property of quarks roughly analogous
to electrical charge. A quark can have any of three possible
color charges, typically referred to as red, green, and blue.
In spectra, these three colors combine to make white light.
Particles have to be made of combinations of quarks whose
color charges add up to white.

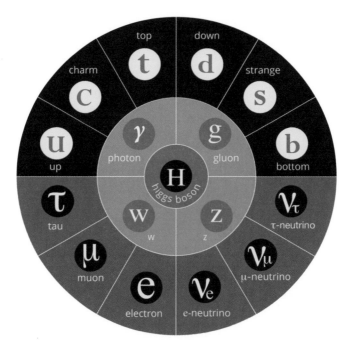

Today's chart has more than 120 elements, but we can make a simpler table,
not of elements but of their component subatomic particles,
all observed experimentally.

■ **Quark flavor** Ⅰ Designates which kind of quark we are talking about. The up and down quarks, for example, have different flavors.

■ **Gluon** Ⅰ A particle that, when exchanged between quarks, generates the force that holds quarks together in particles. Like quarks, gluons carry the color charge.

■ **Strangeness** Ⅰ A property of the strange quark that is roughly analogous to electrical charge. One of its effects is to slow down the decay of particles that contain this particular quark.

■ **Charm** Ⅰ A property with the same effect as strangeness but specific to charm quarks.

■ **Bottom and top** | Properties with the same effect as strangeness and charm but specific to bottom and top quarks.

ARE THERE EVEN MORE LAYERS?

Are quarks really the end of the story about the basic structure of the universe? Have we arrived at the 0's and 1's of the cosmos? The very frontiers of theoretical physics seek to answer this question. Two approaches in that field lead us to the next step, if there is a next step to be taken: string theory and loop quantum gravity.

STRING THEORY | As the name hints, these ideas picture the different quarks as different modes of a vibrating string. These strings are very small: a million-billionth the size of the atomic nucleus. Size isn't the real problem with strings, however; the problem is that for the theories to make mathematical sense, strings have to vibrate in 10 or 26 dimensions.

We live in a world of four dimensions. Think of the last time you agreed to meet up with a friend. Perhaps you said, "Meet me on the 86th floor of the Empire State Building, on the corner of 34st Street and Park Avenue." That's three coordinates (elevation, latitude, and longitude), or three dimensions. But those three numbers wouldn't have been enough to find your friend. You still had to answer the question "When?" Time is our fourth dimension.

To plan a meeting in a string theory universe of 10-dimensional space, you would have to give 10 coordinates. We live in four of them, but the remaining dimensions are too small to notice.

By analogy, if you look at a garden hose from a distance, you easily identify a single dimension: the length of the hose, for example. Seen from afar, then, the hose is a one-dimensional object. Look more closely, however, and you see that the hose is actually three-dimensional. Two of its dimensions, width and depth, are very small compared with its length. In the same way, theorists argue, we will see the extra dimensions of strings only when we can probe them with energies much higher than those available to us today. And the multiple vibrations of these strings in a multidimensional space manifest in our dimensions as the particles we see.

LOOP QUANTUM GRAVITY | Another approach looks not at the structure of quarks but at the structure of space-time itself. In most theories, space and time are simply constant backdrops against which events unfold, like a stage upon which a play is performed. Look at a small enough scale (or, equivalently, at high enough energies), theorists argue, and space and time will become granular, or quantized. At these scales, the structure of space becomes a weave of interlocking loops—like a piece of chain mail. In this scheme, particle interactions do not take place *in* space, but interact *with* space.

String theory and loop quantum theory have two important features in common:

- **Both purport to be the ultimate theory to explain the structure of the universe—what physicists call the "theory of everything."**
- **There is no experimental evidence for either hypothesis.**

Thus we have reached the current limits of our knowledge, both practical and theoretical, about what the universe is made of. And we're left with more questions now than when we started.

WHAT IS

LIFE?

The intricacies of life: forsythia pollen via scanning electron microscope

Fig. 179* to 488.

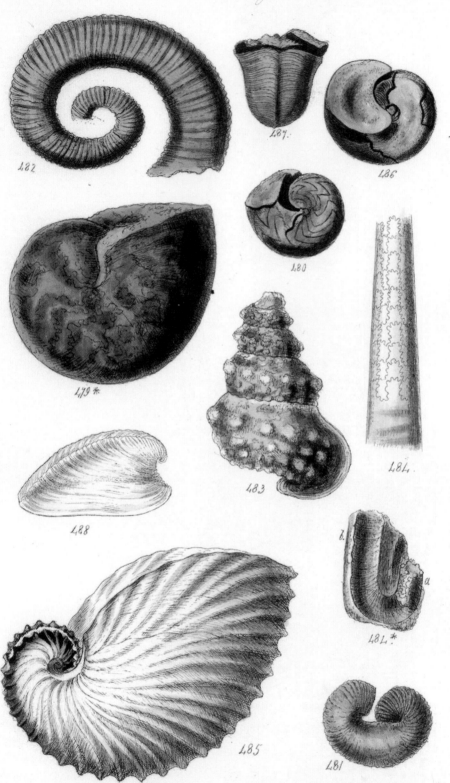

182

187

186

179*

180

184

483

488

485

b.

a

484*

481

6

If we're ever going to find life elsewhere in the universe, we should probably have some understanding of life here on Earth. But let's be honest. A clear definition of life eludes even the most seasoned of biologists. Life, an ever evolving concept, remains on the periphery of scientific understanding.

We can identify three different ways people have tried to define life: (1) definition by list, (2) definition by history, and (3) definition by thermodynamics.

DEFINITION BY LIST | Look in any biology textbook and you are likely to find life defined by a list of properties, with the conclusion that anything possessing all or most of these qualities is alive. The list might include being composed of cells, or a capacity to adjust to the environment, or the ability to reproduce.

Obviously, a list like this is Earth-centric, and might not apply on an exoplanet. In addition, lists inherently find ways of inviting others to point out problems with them.

Graceful curves and spirals characterize these shells and fossils, life-forms painted for a 19th-century handbook.

DEFINITION BY HISTORY | In 1994, NASA convened a panel to explore the definition of life. They decided: Life is a self-sustaining chemical system capable of Darwinian evolution, also known as evolution by natural selection. All living things on Earth are thought to have descended from a first cell that appeared, presumably, in the oceans.

Once again, we have a clear definition of life on Earth, but no way of extending it to worlds beyond.

DEFINITION BY THERMODYNAMICS | The second law of thermodynamics states that an ordered system, when left to itself, will always descend to a state of disorder. Thus an ice cube, a system with high order, will melt into a puddle of liquid water, a state of high disorder.

But living systems are clearly in a state of high order; just imagine what you'd look like if you randomly reassembled all the cells in your body. Still, just as an ice cube can maintain its form in the freezer, itself sustained with electricity, living systems can sustain high order only if they have access to energy. The thermodynamic definition of life, then, is a system maintained in a state of order by a flow of energy.

THE EXPERIMENT THAT CHANGED EVERYTHING

By the end of the 19th century, scientists had discarded many of their old misconceptions about life and embraced new ones. No one believed anymore in things like spontaneous generation—the belief that living organisms can simply arise from nonliving matter. With such ideas a thing of the past, the germ theory of disease was poised to revolutionize medicine. We soon came to learn that life is based on chemistry and, as German biologist Rudolf Virchow declared in his famous pronouncement, "*Omnis cellula e cellula—All* cells come from cells."

And yet one question remained out of reach: How did life itself originate? If all cells come from other cells, where did that first cell come from? A dark and uncharted gulf appeared to separate matter that is living from matter that is not, with no cross bridge in sight. Simply stated, molecules found in living systems are vastly more complex than those in nonliving systems, and the link between them remained inscrutable. Consequently, the question of the origin of life was left to philosophers and especially theologians, and largely ignored by scientists.

All that changed in 1952, with a simple experiment run in a basement laboratory at the University of Chicago. Harold Urey, a Nobel laureate, and Stanley Miller, then Urey's graduate student,

Stanley Miller returns to the 1952 equipment he and his professor, Harold Urey, used to emulate the chemistry of early Earth, in an attempt to generate molecules necessary for life.

set up an apparatus to mimic the conditions on early Earth. A large glass bulb contained water to represent the oceans, electric sparks simulated lightning as a source of energy, and heat was applied, to represent the warmth of the Sun. They filled the rest of the bulb with gases then thought to have been present in Earth's early atmosphere: water vapor, methane, ammonia, and hydrogen, written as H_2O, CH_4, NH_3, and H_2.

After the system ran for a few weeks, Miller and Urey noted that the water turned a murky, brownish maroon color. Chemical analysis revealed that it contained molecules of amino acids, which are the basic building blocks of the complex proteins that run the chemical reactions in living systems. In other words, it appeared that Miller and Urey's primitive apparatus, which had begun with simple molecules that were clearly not a part of living systems, had produced complex molecules characteristic of living systems. At least part of the gap between life and nonlife had been bridged.

Although Miller and Urey's experiment was a historical milestone, it was not perfect. First of all, their best guess for the gases of early Earth was wrong. Instead of methane and ammonia, they should have used nitrogen and carbon dioxide. But, as it turns out, this mistake scarcely matters. Subsequent experiments carried out by generations of researchers using a wide variety of conditions have not only reproduced Miller and Urey's results but also created many kinds of complex molecules, up to and including DNA. Furthermore, amino acids and other organic molecules have been found in meteorites and even in interstellar gas clouds. Nature has ways of producing copious amounts of life's molecular building blocks—in part because those molecules comprise atoms that are the most abundant in the universe.

That gap doesn't seem nearly so wide nowadays as it did a century ago.

Might meteorites have brought the essence of life to our planet?
As this illustration shows, Earth and the Moon experienced showers of space
rocks some four billion years ago. Many meteorites contain amino acids,
key biological building blocks.

DID SPACE ROCKS SEED LIFE ON EARTH?

Scientists have known since the mid-20th century that meteorites—the rocky and metallic remnants of asteroids and comets that survive the descent through Earth's atmosphere—contain amino acids, the building blocks of life. Researchers have studied space rocks that crash-landed on every continent of our planet. But the best ones are found in Antarctica, where a black speck of rock is easily distinguishable from its homogenous white surroundings, undisturbed by civilization. Astrochemists who have studied these meteorites think the extraterrestrial amino acids likely formed billions of years ago during the birth of the solar system, hitched a ride through space inside the rocks, and ended their journey on our planet. Many scientists think Earth was seeded with these enriched space rocks, helping to jump-start life on our planet.

Another benefit of analyzing meteorites from Antarctica, aside from their being easier to spot, is the pristine nature of the surrounding ice in which the rock is embedded. When researchers have compared the amino acids in that ice to those inside the meteorite, they don't match—offering good evidence that the amino acids in the rock did not result from Earth contamination but were truly forged in space.

But how do we know these amino acids formed billions of years ago?

Before 2007, scientists could only study the precious few space rocks that had fallen to Earth and survived erosion by the time they were found. However, NASA's Stardust mission—the first ever spacecraft to return samples from a comet—finally enabled scientists to study the composition of these space rocks as they exist in space. Comets, ancient icy bodies undisturbed by the erosive and corrosive forces on Earth's surface, tell the story of how the solar system formed and what organic molecules may have formed with it. Traceable to the earliest days of our solar system, comets encase organic particles from that time in ice, preserving them for billions of years in their faraway orbits within the Kuiper belt and Oort cloud.

When a comet comes near the Sun, however, the ice evaporates, dislodging ancient dust grains in its orbit, which form part of the comet's tail we see from Earth. The Stardust probe passed through a comet's tail, collected dust samples, and returned to

Meteorites lurk among native rocks on a blue ice moraine in Antarctica. The least contaminated examples of comet and asteroid fragments, these frigid rocks may hold clues about the origins of life on Earth—and beyond.

Earth. And—by now to no one's surprise—researchers found amino acids in the samples.

Nine years later, the European Space Agency's Rosetta mission supported this finding when its instruments detected the amino acid glycine in the tail of another comet.

RNA WORLD

Which came first, the chicken or the egg?

This old conundrum, so often used to puzzle and confuse children, has special relevance in the story of life's origins. The

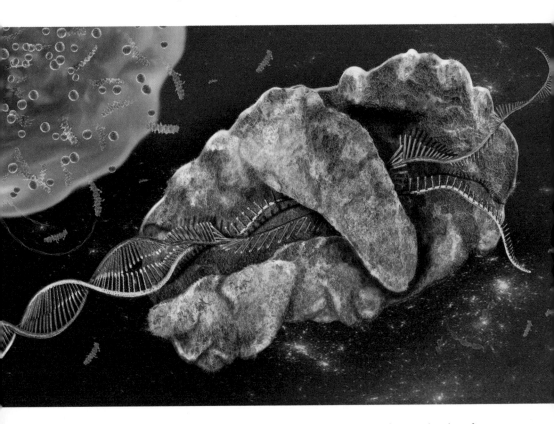

A strand of DNA untwists, undergoing transcription to produce molecules of RNA. This illustration also shows the cell nucleus (upper left), which offers nucleic acid fragments necessary for the process.

chemical reactions that keep an organism alive are governed by complex molecules called enzymes, a kind of protein that enables other specific reactions to occur. The code that governs the production of proteins is contained in the DNA (deoxyribonucleic acid) molecule.

DNA is like a set of instructions contained in the office of a factory. To turn those instructions into a finished product, the information they contain must be carried out to the factory floor, where the actual assembly is done, accomplished for life by a molecule called RNA (ribonucleic acid). The DNA molecule also codes instructions for making RNA.

So here's the problem: To make the enzymes that govern life's chemical properties, we need to have RNA. But to make RNA, we need to decode the instructions to do so contained in the DNA, and decoding is itself a chemical process run by enzymes. So to make the RNA, we need enzymes, but to make the enzymes, we need RNA.

Back to our original question: Which came first, the chicken or the egg?

In the early 1980s, American biochemists Thomas Cech and Sidney Altman discovered a possible way out of the chicken and egg problem (and shared the 1989 Nobel Prize for the discovery). They found that certain types of RNA could act as enzymes in chemical reactions. If one of these types of RNA was made like the ones in the Miller–Urey experiment, it could act as an enzyme and run chemical reactions—but also contain the coding to produce both itself and, eventually, ordinary protein enzymes. In this scenario, these special types of RNA would be the first complex molecules, ultimately leading to the modern cell.

This RNA world, though broadly supported by biochemists, is not the only hypothesis for the origin of life. Some suggest, for example, that claylike minerals may have substituted for enzymes by arranging molecules on their surfaces via electrical charges. Others suggest that the original cell may not have used enzymes at all, but engaged a simple chemistry that could run without them.

Everyone agrees, though: That first cell, no matter how it appeared, changed Earth forever.

NATURAL SELECTION

We know that nature on its own can assemble the basic building blocks of life. The main gap in our knowledge is how a primitive cell—an organism that can carry out chemical interactions and reproduce itself—first appeared from them. Once that original cell appeared, a new process called natural selection took over, providing robust pathways for that original cell to transform over time into the diversity and complexity of life we find on Earth today.

SURVIVAL OF THE FITTEST

Charles Darwin did not write the words "survival of the fittest" in his original work. The phrase was first coined by the English biologist Herbert Spencer after reading Darwin's 1859 *On the Origin of Species*. The phrase ultimately did appear in the fifth edition of Darwin's book, and it has since become the accepted shorthand phrase for the concept of natural selection.

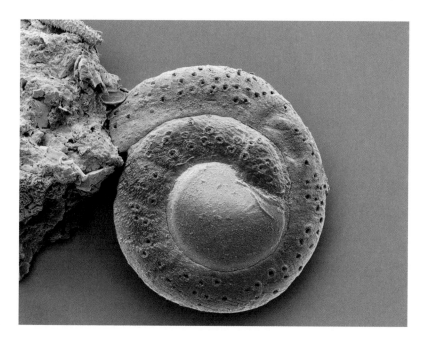

Some 4,000 species of foraminifera—single-celled marine life-forms—inhabit oceans worldwide. Their 500-million-year fossil record reveals change through time, driven by natural selection.

That lonely first cell was capable of taking in molecules from the environment, running chemical reactions, and reproducing itself—perhaps by the RNA methods described previously. Eventually, Earth would be teeming with these primitive cells, feeding on the available environmental resources.

Sooner or later, an agent in the environment—radiation or heat or chemicals—would change one of the molecules in a cell. We call this a mutation. More often than not, mutations cripple the cell so that it cannot reproduce. Occasionally, however, a mutation results in a cell that can exploit the environmental resources more efficiently than its fellow organisms. More important, the cell will be able to reproduce better than others. That mutation is copied and handed down to all the cell's descendants. Eventually, the entire cellular population possesses that same mutated molecule. This is what we mean by natural selection.

Natural selection enables and drives biodiversity. Although storms may blow some cells into Arctic waters, others remain in the tropics, and the different environments favor different mutations. Keep that up for long enough, and eventually we have different species of cells in different places.

Natural selection is a simple, logical process that we can expect to find operating on worlds far different from Earth. This doesn't mean the results will be the same; different environments on exoplanets will certainly produce and favor different kinds of life. A large rocky planet with much higher gravity than Earth might favor shorter, squatter species, for example. A tidally locked planet would encourage wholly different life-forms. With one side in perpetual daylight, always facing the heat of its host star, such a planet would consequently suffer from fierce winds driven by severely unequal heating across the planet's surface—circumstances that might favor a body aerodynamically suited to maneuver and seek nourishment in that grueling environment.

IS COMPLEXITY INEVITABLE?

For the first two and a half billion years of its existence, Earth was singularly uninteresting. An extraterrestrial visitor would have found an ocean world with blue-green photosynthetic gunk along the shorelines. The photosynthetic material consisted of single-celled microbes with a simple, primitive structure. The cells didn't even have a nucleus, only free-floating DNA inside their walls.

About two billion years ago, a large cell swallowed a smaller one, and the two found they were much more successful working together as one than they had been when apart. This mutually beneficial relationship, called symbiosis, catapulted life down the road toward the complexity we see in living organisms today.

The ocean gunk continued to evolve for another billion years or so, before another crucial event. Around 800 million years ago, a group of cells found they could be more successful if they banded together, instituting a kind of division of labor rather than remaining isolated as individuals.

The first multicelled life was simple—and yet the history of its complexity resembles the modern American interstate highway system. It depends on all manner of supporting technologies: Someone has to build and sell cars, someone else has to produce and distribute gasoline, someone has to lay the pavement, and so on.

We know, however, that the highway system did not appear all at once, or out of nowhere. Modern roads often began as game trails and footpaths. In time, they morphed into unpaved wagon roads. Then people like Henry Ford came along and built cars. Lots of cars. And, little by little, roads were paved and filling stations were built. The modern interstate highway system, in all its complexity, arose from a long evolution of many steps and advances—some small, some large. Same goes for life.

INTELLIGENCE & TECHNOLOGY

Technology is an outcome of intelligence. When we employ scientific knowledge toward a specific purpose, and toss in some clever engineers, you get technology. Technology is the invention of the wheel to transport heavy objects, the campfire to cook food, and the smartphone to connect with the rest of our species.

The evolution of human technology required two precursors: complex cells and multicellular life. Each of these took more than a billion years to appear on Earth. If this means these steps are evolutionarily difficult, then intelligence and technology may not be as common in the universe as we might hope, or expect.

On the other hand, it doesn't take much of a brain to produce highly complex behavior. The tiny-headed honeybee, for example, executes sophisticated mathematical waggle dances to communicate the location of distant food sources to its fellow bees. The common octopus, with its primitive brain, can navigate mazes and often escape its enclosure—and let's not forget that they actively and independently control eight appendages.

An Indonesian coconut octopus scuttles along the ocean floor, seashells tucked beneath its tentacles, a hideaway at the ready.

But no useful chart exists to compare the wide variety of brains and intelligence found in Earth's animal kingdom. Depending on how you define intelligence, then, it could have appeared early in the evolutionary record. Even the most primitive behaviors, like spotting and fleeing a predator, confer an evolutionary advantage. But also, if intelligence is so important for survival, then why are we at risk of rendering ourselves extinct thanks to the creations of our own intelligence?

And must intelligence always lead to technology? After all, the dinosaurs ruled the world for more than 200 million years and, as far as we know, never developed campfires, never proposed the theory of general relativity, or never watched Netflix. They most surely would have kept on for longer had not a random asteroid collided with Earth 66 million years ago, ruining their day. We might then imagine countless worlds out there ruled by dinosaurs who didn't have such bad luck.

ORGANELLES

The cell found in humans and other multicelled organisms contains many complex internal structures called organelles, each thought to be the result of a symbiotic event during the evolution of life. The cell's DNA is contained in the organelle called the nucleus, and the cell's energy is generated in organelles called mitochondria. Other organelles serve different functions, all of which the cell requires to remain alive.

SYNTHETIC LIFE

What if we consider the kind of life we know—chemical-based, in an enclosed fluid environment—as just a way station on the journey to something else? What if an endpoint in the evolution of organic life is to create another life-form beyond biology—one that evolves from modern computers? Some scientists and futurists predict this trajectory of human evolution as not only possible, but inevitable. Science fiction authors cleverly refer to this as *Homo siliconensis,* because silicon chips are major ingredient of transistors—which, in turn, are the main working parts of modern computers.

In 1965, the American engineer Gordon Moore, co-founder of the Intel Corporation, made a remarkably prescient prediction based on the rapid advance in computer technology. His idea came to be known as Moore's law. He predicted that the number of transistors on a chip—and therefore every measure of computer performance—will double every two years (a time later reduced to 18 months).

Moore's law isn't a law of nature, like the law of gravitation, but

THE PAPER CLIP UNIVERSE

Oxford University philosopher Nick Bostrom formulated an amusing dystopia related to the singularity. Suppose, he argued, that you design a robot to take materials from the environment and make paper clips. The machine makes more and more effective versions of itself until eventually the entire universe has been turned into paper clips. It's not malevolent. It doesn't hate you. It just needs atoms from your body to make paper clips with.

Or, more positively, as computer pioneer Danny Hillis and colleagues put it in the slogan for their company, Thinking Machines: "We're building a machine that will be proud of us."

Neil deGrasse Tyson ✓
@neiltyson

Odd that our measures of animal intelligence are often tests for what we do best, rather than tests for what they do best.

💬 891   🔁 13.9K   ♡ 48.5K   11:29 AM - Feb 10, 2017

it's an observation that has proved strikingly accurate over the past half century. Transistors have indeed shrunk to smaller and smaller dimensions. Computers, once the size of refrigerators, can now fit into the palms of our hands. Transistors today are so minuscule that they will soon butt up against a fundamental

Romeo, an autonomous programmable humanoid robot, can walk, climb stairs, and grasp objects, and is learning to estimate age and discern emotion based on the faces it views.

limit imposed by the laws of physics. They just can't get much smaller.

The American futurist Ray Kurzweil predicts that we will actually surpass the physical limits to Moore's law and continue to exponentially improve computing technology by transcending the silicon chip. One major contender to fulfill Kurzweil's prophecy is the advance of quantum computing. That's a computer that would harness quantum entanglement to reduce complex algorithms to their simplest solutions in a fraction of the time modern computers require.

Computer engineers around the world are in a kind of arms race to perfect the technology right now. If computers continue on Moore's exponential growth trajectory, then in 20 years, they will be a thousand times more powerful, and in 30 years, a billion times more powerful than they are today.

So, what will happen when the ability of machines mimics and then outperforms human intelligence? More important, what will happen if the machines become self-aware and capable of improving themselves? Would we recognize such machines as being alive? Should we?

This hypothetical situation is referred to as the singularity, a word loosely borrowed from mathematics and astrophysics, meaning the specific point at which an object is not defined or predictable. Singularity is also the name of the center of black holes—a place where the very laws of physics as we understand them break down. Like the technological singularity, it evades any conclusive, evidence-based predictions. Here, only hypotheses reign.

Living examples of artificial systems abound in science fiction. Think of the TV characters named Data (an android) from *Star Trek: The Next Generation* and The Doctor (a hologram) from *Star Trek: Voyager,* or the Terminator robots in the eponymous movie, or the replicants in the cult classic, *Blade Runner.* If we encountered such entities, would we argue that their

Neil deGrasse Tyson ✓
@neiltyson

Seems to me, as long as we don't program emotions into Robots, there's no reason to fear them taking over the world.

💬 985   🔁 2.8K   🤍 4.4K   11:29 PM - Aug 8, 2014

behavior qualifies them as being alive, or would we fixate on the fact that they were manufactured, not born?

To our mind, the likeliest artificial life-form we might encounter is something called a von Neumann probe. Named after the Hungarian-American mathematician John von Neumann, these are small, intelligent probes sent out by an advanced civilization to nearby exoplanets. Upon arrival, the probes begin terraforming the planet and creating the infrastructure that will be needed when members of the advanced civilization that sent them arrive at a later date. The first thing the von Neumann probes do upon landing, by the way, is to make multiple copies of themselves from the natural resources on the planet, to be launched toward ever more distant exoplanets. If you do the math, the number of exoplanets visited by these probes grows exponentially. Thus the von Neumann probes can be thought of as a wave of colonization spreading throughout the Milky Way—a wave that will continue to expand whether or not the original advanced civilization survives.

LIFE OF ANY OTHER KIND

As far as we can surmise, several fundamental principles of nature limit what life on Earth can be. Temperature and time are two of these components, and they are surprisingly interconnected.

With a few rare exceptions, all living things on Earth inhabit environments with temperatures roughly between the boiling and freezing points of water. Take life to lower temperatures,

> **Neil deGrasse Tyson** ✔
> @neiltyson
>
> Any time we're impressed by what a non-human animal does, it's simply because we previously underestimated its intelligence.
>
> 💬 246   🔁 5.9K   ♡ 10.8K   11:12 AM - Aug 24, 2015

and it hibernates or dies. Take life to higher temperatures, and you kill it. In principle, life could exist outside these limits. If it does, it will be unlike anything we've ever imagined.

So what does temperature have to do with time? On average, a chemical reaction that runs at a specified rate at one temperature can be expected to take twice as long if the temperature is lowered by 10°C. This is why food will spoil in a few days if left on your counter but will last for weeks in your refrigerator and months in your freezer. Spoiling, after all, involves unwelcome chemical or biological reactions. On Saturn's moon Titan, temperatures hover at a chilly minus 300°F (about -200°C). That's cold enough to freeze water permanently into bedrock and to liquefy methane gas, creating rain and rivers and lakes of the stuff.

Does life need liquid water? Or just liquid?

These temperatures mean that any metabolic process that takes an organism on Earth a minute to complete would require a couple months on Titan. Leaving open the question if, at such a low temperature, a life-form would require months or years to draw a breath, would we even recognize it as being alive? Or would we discount it as an inanimate object?

At the other end of the scale, high temperatures make

Carried by Cassini, in 2005 the Huygens probe (illustrated here) reached Titan, a moon of Saturn where life-forms may be found. The first successful landing in the outer solar system, the probe relayed data for 72 minutes.

particles move very fast, which means collisions between complex molecules become violent enough to tear them apart. Thus, we don't expect to find living organisms in molten lava.

Temperatures on Earth enable life as we understand it to operate on timescales of seconds and minutes. Think how long it takes to draw a breath or feel a pulse. What we earlier called synthetic life, on the other hand, might operate on a much faster timescale, because it may not be as limited as the fragile, temperature-sensitive, organic life-forms we know on Earth.

EXTREMOPHILES

Though we do not expect life to flourish in molten lava or frigid methane lakes, we do know that some microbes can not only survive in, but prefer, the temperature of boiling water in places like the hot springs of Yellowstone National Park or even the arid, hypersaline, high-altitude salt flats of the Andes Mountains.

Those organisms that thrive in such lethal environments are collectively called extremophiles, which literally means "lovers of extremes." These creatures can flourish in places with extraordinary conditions, such as uncommonly hot or cold temperatures, high acidity or alkalinity, high or low pressure. We've found extremophiles deep in Earth's crust as well as in the deepest and darkest parts of the ocean, where pressures reach more than a thousand times sea level—the equivalent of 15,000 pounds per square inch.

The microscopic tardigrades, for example, affectionately known as water bears or moss piglets, have proven themselves to be some of the most indestructible of all the extremophiles. These eight-legged water-dwelling creatures—creepy yet adorable—may be the hardiest, most unkillable life-form ever discovered. They can survive practically anything you throw at them. They have even endured trips to outer space. In 2007, the European

HYDROTHERMAL VENTS

In 1977, marine geologist Robert Ballard uttered the words, "Wait a minute. What is that?" He was on a research vessel in the Galápagos, looking at pictures conveyed from unmanned submersibles roaming deep underwater. As it turned out, he was witnessing the world's first picture of hydrothermal vents—a discovery that would upend our very understanding of life on Earth.

Hydrothermal vents occur at fissures on the seafloor where Earth's tectonic plates meet. Seawater seeps down into the fissures, mixes with the molten lava below, and spews back out as chemically and minerally enriched water, superheated to about 700°F (370°C). The prevalence and concentration of chemicals such as sulfur and carbon dioxide would be toxic to most animals—and yet it turns out that hydrothermal vents harbor life.

The ocean floor, once thought to be a barren desert of sunless, near-freezing temperatures and thousands of pounds of pressure per square inch, is now known to be a flourishing ecosystem. The bacteria there, adapted to use chemicals instead of sunlight for energy—a process called chemosynthesis—in turn feed neighboring plant and animal species, all with their own unique adaptations to survive the most inhospitable environment ever discovered to harbor life on our planet.

Tube worms thrive in the scalding, oxygen-poor environment of deep-sea vents.

Space Agency strapped tardigrades to the outside of a capsule that was sent into low Earth orbit for 12 days. The tardigrades, exposed to the vacuum of space and extreme cosmic radiation, survived the journey.

Even more remarkable, however, is the tardigrade's ability to survive without water for decades. When humans and most other organisms go without water, the enzymes and DNA in our cells quickly shrivel into dysfunction. A week to 10 days without water, and we are dead. Deprived of water, tardigrades enter into a state

Tolerating extreme conditions such as frozen polar lakes, boiling hot deep-sea vents, and even high doses of radiation, tardigrades have broadened our definition of life on Earth and diversified our search for life-forms on other planets.

Neil deGrasse Tyson ✓
@neiltyson

The pudgy, lovable, mildly creepy, microscopic Tardigrade "WaterBear" would make a most excellent @Macys Thanksgiving Day parade balloon.

💬 869   ↻ 7.6K   ♡ 33.9K   9:47 PM · Nov 22, 2017

of suspended animation, pausing almost all metabolic activity—the deepest form of hibernation known.

Tardigrades have a role to play both in science fiction and on the frontier of science as we develop ways for humans to survive long-term space voyages. All we need to do is unlock their secrets of survival.

The better we understand life on Earth, the more informed our search for extraterrestrial life will be.

ARE WE ALO
UNIVERSE?

Are we alone? Human
nature compels us to
look up and wonder.

NE IN THE

EUROPA

DISCOVER LIFE UNDER THE ICE ⟨ALL OCEAN VIEWS!!!⟩

7

Anyone who grapples with the questions "What is life?" and "Are we alone?" is inevitably handicapped by the limits of our knowledge: The only kind of life we have ever known or studied exists exclusively on Earth. But life on exoplanets may look and function unlike anything ever observed—and to proceed with any search for life out there, we need to acknowledge our tendencies toward shortsightedness.

LIFE LIKE US | There was a time, long before DNA sequencing and other advances in biotechnology, when we bucketed life into two categories: plants and animals. But we have since learned of breathtaking diversity among single-celled and multicellular life thriving on the planet. Still, all known life-forms on Earth—including animals, plants, protists, fungi, archaea, and bacteria—share a basic chemistry. They all structure themselves around a molecular backbone of carbon atoms.

Hence, understandably, people presume all living things must be structured this way—that all life is carbon based, built just like the life-forms on this world.

Might we one day see travel posters luring us to Europa, a moon of Jupiter, whose subsurface liquid ocean could possibly harbor alien life?

Neil deGrasse Tyson ✓
@neiltyson

Note to HOLLYWOOD:

A space alien with no DNA in common with life on Earth
should look more different from life on Earth than
any two life forms on Earth look from each other.

💬 2.8K   ↻ 4K   ♡ 41.9K   3:18 PM · Jun 24, 2020

Hollywood sci-fi films reveal this assumption and this self-favoritism with the humanoid-shaped aliens dominating the genre. But why should aliens have teeth, shoulders, and fingers like human beings? Actually, why should they look like any plant or animal found on Earth? What if life elsewhere in the cosmos were even more different from us than we are from, say, *E. coli* bacteria? What might life look like then?

LIFE NOT LIKE US | Let's explore two ways life not like us could develop.

It could be composed of molecules made from atoms other than carbon. One example of this, popular among science fiction writers, is life based on silicon, rather than carbon.

A CASE FOR LIFE ON EARTH

Carbon is 10 times more abundant in the universe than silicon, its closest chemical cousin. And considering the molecular fertility of carbon, all urges to imagine life based on silicon, although rational, are simply unnecessary. Also, who are we to presume that models for life will be entirely different on other planets? Maybe Earth's life laboratory is literally universal. The laws of physics are. The chemical elements are. The rocks and minerals are. Why should life be an exception to this trend?

An illustration of exoplanet 55 Cancri e, which orbits close to its host star and is tidally locked with it. For these reasons, the entire star-facing surface likely bubbles with molten lava.

Silicon is an appealing carbon substitute because it has an electron structure similar to carbon's. It sits directly below carbon on the periodic table of elements, and so it, too, can bond with four different atoms—a convenient feature for building complex molecules like DNA. But silicon bonds tend to be stronger than carbon bonds, rendering silicon an unlikely candidate to form complex molecules—and, therefore, complex life.

The second way life not as we know it could arise is within a liquid environment other than water. We know of at least one place with non-water lakes: Saturn's largest moon, Titan—the only other world in the solar system known to have flowing liquid on its surface. As previously discussed, its pools of liquid methane and ethane at a temperature of minus 290°F (-180°C) span much of its poles. By comparison, the coldest temperature ever recorded on Earth—in Antarctica—is minus 128.6°F (-89°C).

At the other extreme, there might be an exoplanet lava world where life-forms flourish in a scorching, molten stew. We just don't know what complex chemistry goes on at these extreme temperatures, where something completely unexpected may await our discovery.

Neil deGrasse Tyson ✔
@neiltyson

You'd be inexcusably egocentric to suggest that Earth was the only place in the observable Universe with life - - among the hundred-billion galaxies, each containing a hundred-billion stars orbited by a hundred-billion planets.

Yet how terrifyingly lonely it would be, if true.

💬 2K   ⟲ 12K   ♡ 64.8K   8:01 AM · Jun 24, 2020

LIFE VERY MUCH NOT LIKE US | Until now we have considered only life based on chemical reactions—what we'll call chemical favoritism. Imaginative scientists have speculated, however, about completely different forms of complex structures—for example, an interaction between electric and magnetic fields, or static electrical forces between dust grains in interstellar clouds. What these life-forms might look like, if we could even perceive them with our dull human senses, eludes all but the most open-minded of thinkers.

The amazing spectrum of possible life-forms that could take hold on the innumerable exoplanets in the universe provides a compelling argument that life, intelligent or not, is neither unique to Earth nor unlikely to have come about elsewhere, even if the life-forms on Earth were birthed from an unlikely and rare chain of events.

STRANGE IDEAS

No doubt about it—we humans don't like to think we're alone. From the earliest times we have populated the heavens with living beings, be they gods, demons, or extraterrestrial visitors. There was no limit to where our imaginations might take us.

Only in the last century have we acquired the technology that enables science to examine our ideas about other examples of life.

In the 18th century, some astronomers suggested that the Sun might harbor carbon-based life. Not on the blazing hot surface, of course, but on the solid interior they thought had to be there. Some even imagined that if you aligned your telescope correctly, you could look down through sunspots and see inhabited villages below. This was, after all, before we had acquired knowledge or understanding of the branch of physics called thermodynamics, which tells us that heat from a seething exterior would vaporize any villages in the interior.

As time went on, the Sun lost its luster as a potential home for life—but other strange notions popped up. In 1837, for example, the English parson Thomas Dick published a book with the grandiloquent title *Celestial Scenery, or the Wonders of the Planetary System Displayed, Illustrating the Perfections of Deity and*

In the 1901 H. G. Wells novel and this 1964 film it inspired, humans encounter insectoid Selenites beneath the lunar surface.

WHAT DID LOWELL ACTUALLY SEE?

Of course, we now know there are no canals on Mars. The current consensus is that Lowell was using his telescopes at the limit of their capability. In such a situation, the instrument can produce random dots in the field of vision. The thinking is that in his brain, Lowell connected those dots into a canal network, in much the same way that people connect random patterns in a Rorschach inkblot.

Lowell's 1905 drawing of Martian canals

a Plurality of Worlds; in it, he declared we would actually find people living on the rings of Saturn.

By the early 20th century, the Moon, Mars, and Venus were presumed to host living beings. In 1901, for example, author H. G. Wells, today best known for his earlier novel, *War of the Worlds,* spun a tale of English gentlemen who traveled to the Moon to find a breathable atmosphere and a race he called Selenites. These kinds of beliefs acquired an air of authority when the prominent American astronomer Percival Lowell began publishing books about his observations of Mars. Lowell fantasized the red planet as the home of a dying civilization with a network of canals carrying water from the poles to the equator—one more lost idea about life on Mars.

Today we know the best chances for finding life (most likely microbial) are in the subsurface oceans of outer-planet moons, like Europa, as well as a last-ditch hope that we will find living microbes in aquifers under the Martian surface.

THE SINGLE EXAMPLE

Scientists who study life must labor under a handicap unique to their field. In public, we extol Earth's biodiversity, but behind closed doors we lament that it all traces to a singular origin, a single example of life.

8,141,963,826,080

The number of people Thomas Dick thought lived on the rings of Saturn

The solar system offers more than a hundred spherical objects to compare and contrast with Earth. Among the data, Earth is just one example of a planet. (Incidentally, this is why university geology departments have become so rare in our universities; they've morphed into departments of planetary science.)

Biologists, however, have no such luxury. Every living thing on this planet operates with the same chemistry governed by the DNA molecule, which starkly indicates that we're all descended from a single progenitor cell that appeared in Earth's oceans billions of years ago.

An image of the surface of Mars taken by NASA's Mars Reconnaissance Orbiter shows gullies and rivulets suggesting water, and perhaps life long ago, on the red planet.

Why does this matter? Imagine the only water dweller you had ever seen was a goldfish. Of course you'd assume all water-dwelling creatures were orange vertebrates, liked freshwater, and ate plants and insects. Now imagine that one day you went to the beach for the first time and saw a great white shark, then you found a jellyfish, and then a crab. Everything you thought you knew about water dwellers would need to be reevaluated, and the fields of marine biology and freshwater biology would emerge.

The Allen Telescope Array operated by California's SETI Institute scans the heavens continually for signs of intelligence beyond our solar system.

How might our ideas about life change if we discovered other life-forms out there?

For starters, all life on Earth involves the chemistry of carbon atoms combining in an environment of liquid water. As we shall see in the rest of this chapter, nearly all thinking about extraterrestrial life presumes that whatever we found out there would share this property. Call this the goldfish view of life.

Imagining alien life is like imagining a goldfish for someone who has never seen another water-dwelling creature. One could imagine how life might survive in the water, but to picture and search for a shrimp, a coral, or a 50-ton whale would require more information, time, and especially imagination. A kind of bias, or favoritism, develops, which is a susceptibility of the human mind in the absence of information.

Here are a few sources of bias that the discovery of life elsewhere may (or may not) force us to abandon.

■ **Carbon Favoritism** ❘ Does life have to depend on carbon atoms? Both science fiction writers and serious scientists have thought about life based on other atoms, such as silicon.

■ **Water Favoritism** ❘ Is water the only fluid that can support the processes that led to life? Ammonia and liquid methane represent a couple of other possibilities—and chemists have

PUTTING IT IN PERSPECTIVE

If Earth were a schoolroom globe, the Moon would orbit 30 feet away. Mars, a mile away. The next nearest star, a half-million miles away. If any aliens in our galaxy can traverse such distances, then they're way smarter than we are. Could we even be worth their time to stop and say hello? Could we even perceive their *Hello?* After all, can a worm perceive ours?

suggested others, such as hydrogen sulfide, the molecule that produces the rotten egg smell around thermal pools.

■ **Surface Favoritism** | Must life develop only on the surface of a planet? For many places in our solar system, most of the liquid water is not on the surface but in subsurface oceans, such as those on moons of Jupiter and Saturn. And, for that matter, can life develop and thrive entirely within the atmospheres of the gas giant planets themselves?

■ **Stellar Favoritism** | Can life develop only on planets that circle stars? After all, calculations show that more so-called rogue planets are likely wandering around the Milky Way than planets circling stars. Could life develop independently of a stellar energy source? Could internal heat supplied by radioactivity substitute for sunshine?

■ **Chemical Favoritism** | We ask if life must be based on chemistry. If life requires energy flows, some theoretical calculations suggest that interacting electric and magnetic fields could develop levels of complexity usually associated with living systems.

Since 2012, NASA's Curiosity rover has explored Mars. Findings confirm the basic elements of organic chemistry, suggesting that the planet could have supported life some three billion years ago.

Early Earth's surface was volcanic and lifeless, struck often by comets
and meteors that delivered the basic ingredients of life from elsewhere
in the solar system.

In sequence, questioning each of those favoritisms opens new
and increasingly improbable possibilities for life. Where do you
get off the train?

SEARCHING FOR INTELLIGENCE

If you're going to mount a major search program, it's helpful to
know precisely what you're looking for.

People tend to conflate the search for extraterrestrial life with
the search for extraterrestrial intelligence (SETI). So let's begin with

If a football field were a timeline of cosmic history, cavemen to now spans the thickness of a blade of grass in the end zone

a thought experiment. How would our planet have looked to an extraterrestrial visitor at various points in its history?

For the first half billion years or so, Earth would have appeared as a hot, airless ball floating in space, devoid of life, let alone intelligence.

For the next couple of billion years, Earth would have been a green pond scum world, with relatively simple microbes floating in the oceans and deriving their energy from sunlight. This world would contain life, but obviously not yet what we would call intelligence.

Sometime in the last few hundred million years, our visitor would start finding more complex life-forms. When the threshold to intelligent life is crossed depends on what you consider to be intelligent—Worms? Fish? Dinosaurs? Primates? House cats?

Instead of getting lost in murky debates about the definition of intelligence, let's look at how we search for life on exoplanets, and compare it to how we search for intelligence.

Basically, we use spectroscopy to search for what astrobiologists call biosignatures—molecules in a planet's atmosphere that living organisms can produce. Examples of those molecules include oxygen, derived from photosynthesis, and methane, produced by anaerobic microbes. But there's one problem with this approach: The molecules in question can also be produced by standard chemical and mineralogical processes. We know, for example, that ultraviolet light from the Sun can break up water molecules in the atmosphere, producing molecular oxygen without life.

Stromatolites—rocky structures, such as these in Australia, formed from the secretions of primitive microbes—are now rare but were once the predominant life-form on Earth, some 3.5 billion years ago.

Currently, our only way to detect intelligence in the universe is to search for deliberate or inadvertent electromagnetic signals sent out from an exoplanet. But that means we are defining intelligence as the ability to build radio telescopes. It also means that the entire sweep of human history, from *Homo habilis* two million years ago to the 19th century, would be invisible to extraterrestrial observers using our own definition of intelligent life. Multicelled, or complex, life appeared 550 million years ago. But intelligence defined as the ability to broadcast radio signals represents a tiny fraction—about 0.00002 percent—of the history of complex life on Earth.

Is it fair, then, to assume intelligent life doesn't exist beyond Earth, based on our limited data?

Even though we've sent a flotilla of space probes to Mars and have rovers collecting data on the Martian surface even as you read this, scientists continue to debate about whether or not there is microbial life on the planet. In other words, given what we know now, we might well be living in a galaxy full of green pond scum planets, with perhaps a few dinosaur planets thrown in, but with not a one of them sending us radio signals, or at least any signals we can detect.

THE DRAKE EQUATION

The American astronomer Frank Drake formulated his famous equation in the early 1960s, and it has dominated discussions of the search for extraterrestrial intelligence ever since. The equation allows us to estimate the number of advanced technological civilizations trying to communicate with us right now. It's written as follows:

$$N = R f_p n_e f_l f_i f_c L$$

Where the symbols represent:

N = number of extraterrestrial civilizations
    trying to communicate with us right now

R = rate of star formation per year

f_p = fraction of those stars with planets

n_e = number of Earth-type planets in a
    planetary system

f_l = fraction of those planets that will develop life

f_i = fraction of those that develop intelligence,
    given that life has developed

f_c = fraction of those with a technological civilization
    capable of sending signals

L = length of time that signals will be sent

As we work our way from left to right in the equation, the first three terms involve fairly solid astrophysics. The next three terms involve evolutionary biology and become more and more fuzzy as we move to the right. Finally, assigning a number to the last term would require results from a field we could call exosociology—or the study of interactions between Earthlings and an extraterrestrial civilization.

The uncertainties in the last half of the Drake equation make the difference between a prediction of $N = 1$ (that is, there is only one advanced civilization, and we are alone in the galaxy) to $N =$ millions (or there is a Galactic Club out there that we can join). The popular media, of course, embrace the latter estimates, which leads to cultural expressions like the bar scene in *Star Wars,* in which dozens of vaguely humanoid extraterrestrials mingle together, sipping intergalactic beverages and enjoying jazz music.

But just for fun, let's plug some numbers into the Drake equation and see what kind of results we can generate. Roughly 10 brand-new stars are created in the galaxy each year, so let's take that as the star formation rate: $R = 10$. We're finding planets everywhere, so let's say that at least half of those stars have planets: $f_p = 0.5$. As far as Earthlike is concerned, our solar system harbors about

The cosmic bar in *Star Wars* is a perfect example of exosociology,
a mingling of intelligent life-forms from throughout the galaxy.

100 planets, large moons, and large asteroids, of which only one
(Earth) is Earthlike. Let's assume this is typical and set n_e = 0.01.

Life developed quickly on Earth once conditions were ripe,
so set f_l = 1. For lack of any better evidence, also set f_i = 1. We
discuss the possible inevitability of technology later in this
chapter, but let's anticipate our results there and set f_c = 0.1.

And that brings us to the squishiest aspect of the Drake equation: how long a civilization will broadcast. Originally, physical
scientists chose geological timescales and set L equal to millions
of years; this is where the Galactic Club came from. On the other
hand, humans started broadcasting with radio waves around
1900, but are now substituting cable and satellite communications for broadcasting. Before long, our signals will no longer be

leaking into space. For us, then, L will probably turn out to be about 100 years.

So, depending on what you choose for L—a value we can barely model after our own human civilization—you can get anything from $N = 1$ to $N =$ millions. Regard this as another way to express our ignorance of our place in the universe.

IS TECHNOLOGY INEVITABLE?

The story of life on Earth describes increasing complexity, from microbes to multicelled organisms to the development of technology. But is this development inevitable? Must life, in other words, always lead to intelligence? And must intelligence always lead to technology?

One thing we have learned in our search for exoplanets is that there is a myriad of worlds out there—so diverse that any world you could imagine probably exists somewhere. Molten lava ocean worlds? Probably. A planet made of solid diamond? Why not? A pond scum planet that would eventually house a space-faring species? Well, for billions of years, single-celled organisms inhabited a pond scum planet called Earth. The onset of multi-celled life about 550 million years ago correlates with striking environmental changes: Glaciers retreated and oxygen levels in the atmosphere rose. On a planet without those extraordinary upheavals, the green pond scum might have continued unabated to this day, never leading to intelligent life.

It's easy to see how traits associated with intelligence—knowing where to find food and how to avoid predators, for example—would confer an evolutionary advantage once complex life-forms appeared. But do they necessarily lead to technology?

Again, the history of life on our own planet can point us toward an answer. For more than 200 million years, dinosaurs dominated Earth. In their world, the top evolutionary advan-

THE WOW SIGNAL

On August 15, 1977, Ohio State University's radio telescope, called Big Ear, detected a signal whose origin remains clouded in mystery. It was a strong signal, lasting for a full 72 seconds before Earth's rotation carried the telescope to a position where the source of the signal was no longer visible to it. Astrophysicist Jerry Ehman found the record of the reception when going through a printout of the data a few days later. It was so singular, and so much in line with what we'd expect from an extraterrestrial signal, that he wrote "WOW" on the paper in red ink, forever giving the signal its name.

Well, it's been more than 40 years since WOW, and despite repeated searches, it has never been seen again. Bigger radio telescopes operating in 1977 didn't see the signal at all, nor did some of the detectors on the Big Ear itself. Possible explanations have been proposed: It was an Earth signal reflected off orbiting space junk, or it was radiation from a newly discovered comet. But none of these explanations has achieved widespread acceptance.

Space aliens, anyone?

Perhaps it's best to take the advice of Jerry Ehman who, 20 years after his discovery, warned us not to draw "vast conclusions from half-vast data."

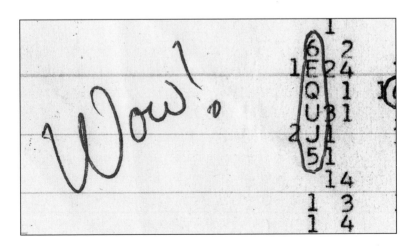

A radio detection in 1977 suggested intelligence out there, but it was never again replicated.

tages were conferred by traits like size and speed. *Tyrannosaurus rex* didn't need tools to do well in its environment, so tools never appeared. Not until about two million years ago did our ancestor *Homo habilis* start down the road to technology by shaping rocks into crude tools. Had the dinosaurs not been driven to extinction by the impact of an asteroid, then mammals (including human beings) might never have become the dominant life-forms. In this case, Earth might have remained a dinosaur planet—a planet with life, perhaps even intelligent life, depending on one's definition—but no technology.

SETI: SEARCH FOR EXTRATERRESTRIAL INTELLIGENCE

The most romanticized endeavor of any space program in popular culture, overwhelmingly depicted on the big screen and in novels, is the quest to discover an advanced technological civilization with which we could converse. Looking for such a civilization is the goal of a program called SETI, the search for extraterrestrial intelligence.

SETI began in 1959, with an article in the prestigious British science journal *Nature* remarking that our emerging ability to build radio telescopes meant that if anyone out there was sending signals, we would be able to detect them. If the phone was ringing, in other words, we could now pick it up. This paper led to the conference where the Drake equation was first written down, and the long, though sporadic, search for those signals began.

The problems SETI researchers face can be summarized by asking two questions: Where do you look? And what do you look for?

The "where" question is fairly easy to answer. We should start searching for signals from planets circling Sunlike stars in our immediate neighborhood—a search for what we've called life like us. The "what" question is more difficult, because there are

Neil deGrasse Tyson ✔
@neiltyson

Imagine a world where Nations find the search for life in the Universe more interesting than the taking of life on Earth.

💬 566   🔁 17.7K   ♡ 22.4K   1:19 PM · Sep 30, 2015

a huge number of frequencies that an alien civilization might choose to employ. The original *Nature* paper, for example, suggested that signals would be sent at a microwave frequency corresponding to a particular phenomenon that takes place in hydrogen molecules, because this was the most common radiation generated by the most common element in the universe. Over the years, other special frequencies have been proposed, but all of them have involved somewhat arcane arguments about how extraterrestrials might view the universe. In the end, the main SETI program settled down to carrying on a daunting, all-sky search for radio frequencies.

The insatiable need for computing power to perform this task prompted astrophysicists at the University of California at Berkeley to start a program in the late 1990s called SETI@home—a project that allows people to become citizen scientists and use their personal computers to analyze SETI data. The Berkeley team sends data packets that can be analyzed by personal computers when they are otherwise idle. You can still walk into offices around the world and see idle computers crunching through SETI data.

Despite the millions of deputized computers, and despite the number of engaged telescopes over the years, no evidence for the existence of an advanced technological civilization in the Milky Way galaxy has emerged. But the search must continue. SETI is one of the few experiments in science whose results would be significant, no matter which way it turns out.

An artist's illustration of the TRAPPIST-1 system, which includes an ultracool dwarf star (far left) orbited by Earthlike planets—candidates for life 40 light-years away

Jill Tarter, chair emeritus at the SETI Institute, put it this way on a *StarTalk* episode: "Take the volume of space and frequencies and time that we might have to search to find extraterrestrial intelligence, and set that volume equal to the volume of Earth's oceans. How much have we sampled in the last 50 years? One 12-ounce glass. So, if you scoop out one glass from an ocean and don't see any fish, can you claim that there are no fish in the ocean? You would be inexcusably short-sighted to do so."

CONTINUOUSLY HABITABLE ZONE

If Earth were somewhat closer to the Sun, we could have wound up like Venus—a scorching, waterless desert. If Earth were a bit farther away, we might now be frozen solid. That delineates an orbital band around our Sun, within which our planet has evolved to support life. We can use that concept to describe the neighborhood around any star where oceans of liquid water could have survived on an orbiting planet's surface for billions of years—the time required to develop complex life. This region is called the continuously habitable zone (CHZ) of a star, where conditions are not too hot, not too cold, but just right to support life—earning the nickname Goldilocks zone.

Every star has a CHZ. It hugs a smaller star more closely than a larger one. And, of course, all sorts of bells and whistles go into the actual characterization of a CHZ—the composition of a planet's atmosphere, for example, and the planet's gravity, which governs whether or not molecules escape into space. No matter how you concoct the perfect Goldilocks zone, though, this concept leads us to seek out not just any kind of life, but life as we know it.

The implicit and explicit thought behind the CHZ is that life requires surface oceans to arise. We have this built-in bias, of course, because that's exactly where life developed on Earth. But planetary surfaces are not the only place we find oceans. The smallish moon of Jupiter called Europa has more water beneath its icy surface than is found in all five oceans on Earth. Thus, if we really want to follow the water in our quest for life elsewhere, those subsurface oceans within our own solar system—oceans that lie well outside of the Sun's CHZ—should be on the list, too.

Nevertheless, the search for life and advanced civilizations elsewhere in the universe has focused on Earth-size exoplanets that lie within their stars' Goldilocks zones. For a while, headlines blared that a possible abode of life had been found every time this sort of planet was discovered. The world's newspapers went positively ballistic in 2016, when astronomers announced the presence of seven Earth-size planets orbiting the star TRAPPIST-1, with three nested cozily in the CHZ.

There is no doubt that the CHZ is a good place to look for life like us—that is, life based on molecules containing carbon interacting in liquid water. The danger is if we perpetuate our favoritisms in our search for life, we may blind ourselves from discovering other kinds of life.

Enrico Fermi sits at the controls of the synchrocyclotron,
one of the earliest particle accelerators, in 1951.

THE FERMI PARADOX

In 1950, physicist Enrico Fermi and a group of colleagues were walking to lunch at Los Alamos National Laboratory in New Mexico. They were chatting about a spate of UFO sightings in the neighborhood, and the conversation naturally turned to the question of whether there really were extraterrestrial civilizations out there. Later, over lunch, Fermi asked a simple question for which we still don't have an answer: Where is everybody?

From the Dad-Joke Vault...

Q: What do you call embryotic Space Aliens?

A: Eggstra-terrestrials.

To see the importance of this question, you have to under-stand a little about Fermi and a little about how people viewed the universe in 1950. Fermi, a Nobel laureate, was the man who had been in charge of building the world's first nuclear reactor. He was also well known for devising quick ways of estimating the answers to seemingly intractable questions, such as "How many extraterrestrial civilizations are out there?"

Although we cannot know for sure, we can make a pretty good guess at what was going through Fermi's mind before he asked his question. A full decade before Drake proposed his famous equation, Fermi had probably made a quick estimate of the num-ber of planets in the Milky Way capable of evolving advanced life. He also would have made a quick estimate of how long it would take an advanced civilization to colonize the entire galaxy.

At that point, he'd realized that (1) there could be lots of extra-terrestrial civilizations out there, and (2) a spacefaring race would take only a few hundred thousand years—a blink of an eye in astronomical time—to colonize the entire galaxy. The results of these calculations raise immediate questions: If the galaxy is really full of advanced civilizations, where are they? Why haven't they contacted us? And if the colonization time is really that short, why haven't the extraterrestrials shown up on our doorstep? The simplest answer is that they aren't there. They aren't anywhere. This conundrum has become known as the Fermi paradox.

Over the years, a number of intriguing responses to Fermi's question and its corollaries have been proposed. Here we offer the top three.

■ **The Zoo Hypothesis** | They're really out there, but for some reason they've decided not to interfere in our development. An example of this hypothesis is the "Prime Directive" in the *Star Trek* series, which forbids contact with primitive societies, lest such contact disrupt their biological or cultural evolution. In this scheme, Earth is seen as something like a zoo or a nature preserve.

■ **The Rare Earth Hypothesis** | The sequence of events that led to the evolution of intelligent life on Earth is so contingent, so unlikely to repeat, that Earth hosts the only advanced civilization in the galaxy. In this scheme, extraterrestrials aren't there because they never evolved. This scenario is popular with religions that favor Earth as God's unique and special place in the universe.

■ **The Doomsday Scenario** | To win the evolutionary struggle, life-forms must use aggression. But when an aggressive organism acquires modern science and technology, this is precisely the power with which that organism can wipe itself out. In this scheme, extraterrestrials destroyed themselves before they start communicating or colonizing.

So . . . where is everybody? Again, we just don't know.

RANKING CIVILIZATIONS

In 1963, the Russian astrophysicist Nikolai Kardashev (not to be confused with the more famous Kardashians) was part of the

THE DYSON SPHERE

Nearly all life on Earth depends on energy from the Sun, but Earth receives only a tiny fraction of the light the star emits. Most of the energy the Sun emits goes off into space, which is true for any star in the sky. After all, the only reason we can see stars at all is that their light has escaped from the region around them. From the point of view of an advanced civilization, then, most of the energy a star generates is wasted.

In 1960, American physicist Freeman Dyson argued that a truly advanced technological civilization would tap into this energy source by building large solar collectors that catch the light before it leaves the region around the star. Fully realized, these collectors completely enclose the star, collecting all of its emitted energy. This structure, which has come to be called a Dyson sphere, would require a truly advanced technology to build. Just ask yourself where you'd get the material to build such a massive structure. It may require that you mine all the natural resources of all planets for the project.

To explain unusual variations in its brightness, some suggest that KIC 8462852, also called Tabby's Star (illustrated here), is surrounded by a partially built Dyson sphere. Most scientists prefer more natural explanations.

KEEPING UP WITH THE KARDASHEV SCALE

1 TWh (terawatt hour, equal to a billion kilowatt hours) = estimated energy production of Earth in 1890

18 TWh = estimated total energy production of Earth today

108 TWh = estimated total energy production of Earth needed to move up to Type 1 on the Kardashev scale

In other words, we need to sextuple our energy production in order to move up to Type 1 on the Kardashev scale.

team that conducted the first SETI in the Soviet Union. As part of his work, he classified civilizations far more advanced than we are—the kind of civilizations everyone expects to encounter out there—into what became known as the Kardashev scale.

In its current incarnation, the Kardashev scale ranks civilizations based on their ability to generate and use energy and puts them into three categories:

Type 1 | Can generate or control all of the energy available within its planet.

Type 2 | Can generate or control all of the energy generated by its host star.

Type 3 | Can generate or control all of the energy generated by all the stars in its galaxy.

In this scheme, the human race hasn't even gotten to Type 1. Extrapolation schemes put us above zero but nowhere near 1 on the Kardashev scale today. There has been progress, though. Some calculations suggest, for example, that when *Homo habilis* started using stone tools a couple of million years ago, we were at 0.1, so we have definitely been moving up.

Some informed futurists think we may be only a few hundred years from achieving full Type 1 status; this would surely include energy generation via nuclear fusion, which we don't yet have.

A Type 2 civilization would certainly have spaceflight and may have built a Dyson sphere, which traps all the energy emitted by the host star and feeds it back to the planet, serving the energy-hungry needs of that Type 2 civilization. This would take place, at best, thousands of years into the human future. A Type 3 civilization is difficult to even imagine. A galaxy composed entirely of Dyson spheres around its stars might qualify, as would a galaxy completely enclosed within a Dyson sphere.

Makes you wonder. If Type 2 or 3 civilizations exist, would their hypothesized Dyson spheres block any signals we might otherwise receive?

As we find new exoplanets, artists combine fact with fancy to evoke possible landscapes throughout the galaxy. Here, a hypothetical icy moon orbits a known gas giant exoplanet 140 light-years away that orbits a Sunlike star.

HOW DID IT

Observations, calculations, and visualizations inform the question of how it all began.

ALL BEGIN?

The most interesting events in the history of the universe happened during the first thousandth of a second. To go that far back in time and understand the origins of the universe, we need to bring together two seemingly disparate branches of science: cosmology and elementary particle physics. That's a curious tryst between the study of the universe, the biggest thing we know, and the study of subatomic particles, the smallest things we know.

Our cosmos has been expanding and cooling for almost 14 billion years. Early in its life, the universe was smaller and hotter, and its constituents collided with one another much more violently than they do in our own relatively frigid era. Complex structures like molecules, atoms, and even elementary particles could not stand the stress of these collisions; only their smaller, simpler components could. The early history of the universe is the evolution of these components.

We've already talked about one of these evolutionary milestones. About 380,000 years after the Big Bang, the universe

A computer simulation centered on a supercluster of galaxies reveals the filamentary large-scale structure of the universe.

cooled enough for atoms to survive the collisions. Until this time, ordinary matter had existed as a plasma, with unfettered negatively charged electrons and positively charged nuclei wandering free, absorbing and re-emitting electromagnetic radiation. The formation of whole atoms neutralized the plasma, cleared the decks, and freed the radiation, which eventually became the cosmic microwave background. At the same time, ordinary matter fell into the gravitational wells shaped by dark matter. The emergence of galaxies, stars, and human intelligence all followed from this event.

The evolution of the expanding universe in a single view, from
the Big Bang (far left) through the formation of atoms, then stars and galaxies,
ending with their deaths

Rewind the video farther to witness a similar transition just three minutes after the Big Bang, when the universe had cooled enough for the nuclei of atoms to survive. Before this time, ordinary matter existed in the form of distinct protons and electrons, but no more complex structures could survive. Even here, when the universe was a mere three minutes old, matter took a familiar form, comprising the same particles we see around us today. And the forces that prevailed in this early chapter of the universe were the same as those we see in action today. For departures from this familiarity, we must go much farther back in time—to when the universe was less than a thousandth of a second old.

THE WAY THINGS ARE

If we're going to talk about how the universe evolved to its present state, we should have a clear idea of what that state is. Ordinary matter in the universe was built from a few elementary particles. Now think of them as the bricks from which a universe is constructed. Some of these bricks are made of baryons—particles normally found in the nuclei of atoms, like protons and neutrons, each made from quarks. As previously discussed, we know of six types of quarks that pair up into what are called flavors: up and down, charm and strange, and bottom and top.

But there is another class of cosmic bricks called leptons, of which we also know six kinds. The most familiar is the electron. But there are two more, called the mu and the tau, which are like the electron but heavier. And for each of these particles, there's a corresponding neutrino, carrying zero electric charge and hardly any mass. Although these exotic leptons may seem strange, all six of them are produced and studied copiously in modern particle accelerators—the huge machines that allow us to generate subatomic phenomena thought to replicate those at the beginning of time.

Neil deGrasse Tyson ✅
@neiltyson

Scientists are simply adults who retained and nurtured their native curiosity from childhood.

💬 825   🔁 13.1K   ♡ 67.3K   12:40 PM - Apr 14, 2018

So the bricks from which all ordinary matter is built amount to six quarks and six leptons.

You read that correctly. Just 12 particles construct the known universe.

How do these particles interact with one another? What forces bind them together or tear them apart? In fact, four forces operate today: the mortar to our bricks of the universe. Two are familiar in everyday life: gravity and electromagnetism. Two

Chemist Marie Skłodowska Curie explored radioactivity. Her findings later helped explain the decay of subatomic particles.

others operate within the nucleus of atoms and are less familiar: the strong and weak forces. Because they aren't as common, let's take a moment to introduce them.

We'll start with the strong force. Protons in the nucleus of an atom all have the same charge, so they want to fly apart due to electromagnetic repulsion. We learned this in middle school: "Like charges repel." But somehow all the protons are content as a clam, huddled together in one place. To overcome the repulsion, there must be a countervailing force holding the nucleus together. Enter the strong force, which became a central object of study in 20th-century physics.

Still, despite the strong force holding them all together, many nuclei and elementary particles undergo radioactive decay—a term coined by none other than two-time Nobel Prize–winning Polish chemist Marie Curie. Radioactive decay happens when the weak force triggers the nucleus of an atom to lose energy by way of radiation.

So there you have it: six quarks, six leptons, and four forces, manifest after the universe's first thousandth of a second. But what happened before then? What pre-endowed the cosmos to give us this particular universe?

QUANTUM MECHANICS 101

To continue our journey backward in time, we need to enter the atom itself. A strange place, described by one of the great theories of physics called quantum mechanics. Quantum is the Latin word for "heap or bundle," and mechanics is the old name for

the science of motion. Thus, quantum mechanics is the study of the motion of things that come in bundles, as they do inside the atom. It's an entirely different world from the Newtonian mechanics of everyday life, so let's take a brief side trip into this quantum world.

For our purposes, we will talk about only one aspect of quantum mechanics: the Heisenberg uncertainty principle, named after German physicist Werner Heisenberg. A version of the principle relevant to our discussion says that the more precisely you know how much energy a system has, the less precisely you know when in time it possesses that energy.

Remember the children's fairy tale, Cinderella? She could dress up and go to the ball, transformed, and no one would be the wiser as long as she got home by midnight. In the same way, in the quantum world, an extra particle can pop into existence—so long as it disappears fast enough, where fast enough is defined by the uncertainty principle. Particles that appear and disappear in this way are considered virtual.

A proton sitting by itself can, for a short time, become both a proton and another particle, like Cinderella at the ball. The mass of the extra particle, after all, is just another form of energy (remember $E = mc^2$), so the extra particle can hide behind Heisenberg's energy uncertainty. So long as that extra particle disappears in a short enough time, the uncertainty principle tells us we will never be able to tell the difference between a system with that particle and one without it.

In the 1930s, Japanese physicist Hideki Yukawa realized the particle that emits a virtual particle need not be the same one that absorbs it. He found that when two ordinary particles exchanged a virtual particle, there was a force acting between them. Think of it like two ice-skaters. When pushed on slippery ice, they easily surrender to the force acting on them and slide in the direction of the push. If one ice-skater throws a heavy ball at the other, the release of mass will cause the skater throwing the ball to recoil, while the one catching the ball will slide backward from its momentum.

In other words, an exchange of undetectable virtual particles in the quantum world generates forces. Furthermore, the type of virtual particle exchanged defines the force.

In fact, the exchanged virtual particle can be identified for three of the four forces. The strong force comes from the exchange of particles called gluons, sensibly named because they enable elementary particles to stick together. The electromagnetic force comes from the photon, while the weak force comes from particles called vector bosons.

QUANTUM MECHANICS IS HARD

Known for his pioneering work in quantum mechanics, the American theoretical physicist and Nobel laureate Richard Feynman once admitted, "I think I can safely say that nobody understands quantum mechanics." So if these concepts feel especially tricky, you're in good company. But the very pursuit of understanding them grants insight to how our universe came to be. And who could argue with the Austrian-American physicist Victor Weisskopf, who once remarked, "There are only two things that make life worth living: Mozart and quantum mechanics"?

That's three of the four forces: strong, electromagnetic, and weak. At the moment, gravity cannot be described this way—a fact that, as we shall see, remains one of the greatest frontiers in theoretical physics.

SIMPLIFICATION & UNIFICATION

When the universe was young, it was hot and simple. And when the universe was younger than that, it was hotter and simpler still. The highly ordered structure of the atoms that formed much later, when the universe was 380,000 years old, was much more complex than the random plasma sea of charged particles from which it evolved. In the same way, the nuclei that formed when the universe was three minutes old were more complex than the jumble of elementary particles that preceded them.

Like dismantling a building into its structural supports and constituent bricks, we can trace the history of the universe back toward its foundations. But another simplification in the quantum world is possible, for which there is no analogue in everyday experience. It involves the four fundamental forces—what we have called the mortar between the bricks of the universe: strong, electromagnetic, weak, and gravity.

Let's start by asking a strange question: How many forces do you need to build a universe? In a universe with no forces, nothing would happen, and that's clearly not the universe we live in. We need at least one force to build a universe, but we really don't need more. In the simplest possible universe, then, there will be only one force. If going backward in time really corresponds with moving toward greater simplicity, there will have to be transitions

This artwork portrays the Big Bang as an explosion—an expansion of matter in space—but the Big Bang was actually an expansion *of* space, not *in* space—a challenging concept to portray.

Neil deGrasse Tyson ✓
@neiltyson

In five billion years, our Milky Way begins to collide with the Andromeda galaxy. But not to worry. The Sun burns Earth to a crisp long before then.

💬 5.4K   ↻ 17.9K   ♡ 110.2K   10:07 PM · Sep 10, 2020

that reduce the number of acting forces, ultimately down to one. We refer to this yet-to-be-achieved idea as the unification of forces—called the unified field theory in Einstein's day.

In the quantum world, forces exist because virtual particles pass between actual particles; we used the analogy of two ice-skaters exchanging a heavy ball to generate a force. Let's extend that analogy and imagine two pairs of ice-skaters gliding toward each other. Perhaps they're in an outdoor ice rink on a cold winter's day. A skater from one pair wields a bucket of water mixed with antifreeze, while a skater from the other pair wields a bucket filled with water that is frozen.

If the skaters now move toward each other and hurl the contents of their assigned buckets toward their partner, we'll see one skater splashed with liquid and the other catching a hunk of solid ice. We would, in other words, see two different phenomena operating in the ice rink. If we repeated the experiment in the summer, however, the bucket of ice would have melted, and we'd see only one kind of phenomenon: an exchange of liquid. We would realize we saw two different phenomena the first time only because the temperature was low. Higher temperatures were required to reveal the two phenomena as the same.

By analogy, as we move backward in time through the first thousandth of a second, the increase in temperature unifies the forces. They begin to behave in the same way as one another. In fact, the farther back in time we go, the fewer distinct forces there were in the universe.

And with this, we are ready to begin our journey into the heart of creation.

QUARK CONFINEMENT

The next significant event that we encounter as we move backward in time happens at 10 microseconds, or 10^{-5} seconds, after the Big Bang. At this moment, free-roaming quarks find companion quarks and settle down into particles, including some we know and love. There were two families of them: three-quark combos formed particles like protons and neutrons, while quark–antiquark pairs formed mesons.

At first glance, this event looks like the formation of nuclei and atoms that occurred much later. There is, however, a crucial difference. The force between quarks is generated by an exchange of gluons, and this force is different from the familiar electromagnetic and gravitational forces in one important respect: As noted earlier, the force between quarks gets stronger, not weaker, the farther apart they are. The farther apart you stretch a rubber band, the harder you have to pull to stretch it more. In the same way, the farther apart two quarks are, the

WRITING SMALL NUMBERS

As we go back in time toward the Big Bang, we talk about shorter and shorter time periods, using numbers that are expressed in what is called powers of 10 notation. A number written as 10^{-3} can also be written as a 1 with the decimal point shifted three places to the left. Thus, a thousandth of a second would be written as 10^{-3} second or 0.001 second. A microsecond (a millionth of a second) would be written 10^{-6} or 0.000001 second, and so on.

During a planned shutdown at the CERN Large Hadron Collider,
an operator installs parts for its new Compact Muon Solenoid detector,
where particle collisions occur and are observed.

harder you have to pull—the more energy you must invest—to separate them.

Now imagine that you have a proton, and you reach inside and grab one of its three quarks. Your goal is to pull out the quark. At first it's easy going—the force holding the quark is small and easily overcome. As you pull your quark farther away, though, the effort becomes harder and harder. Eventually you've pumped so much energy into the system that you've created two more quarks in the form of a meson. At that point, it was easier for nature to convert that energy to quark mass, invoking $E = mc^2$, than to increase the separation farther.

The lesson? No matter how hard you hit a particle, you will never produce a free quark.

THE UNIFICATION OF FORCES

So far, we've moved back through three checkpoints in time: 380,000 years, three minutes, and 10^{-5} seconds after the Big Bang.

Each moment marks a change in the nature of matter. Complex structures like atoms, starting at 380,000 years, ultimately arose from a sea of quarks and leptons, existing before 10^{-5} seconds after the Big Bang. For all this time, however, there has been no change in the fundamental forces involved. Only the bricks of our universe have changed. Our mortar has remained. Even with matter broken down into its fundamental constituents, it still interacts through the same four forces we experience today.

But all that's about to change.

As we continue turning back the clock, the next fundamental change we encounter occurs 10^{-10} seconds (a 10th of a nanosecond) after the Big Bang. It's right around then that forces begin to unify. Before 10^{-10} seconds, there were only three forces acting—the strong force, gravity, and a unified force we can call the electroweak. Only after this time did the electromagnetic and weak forces separate with two distinct identities. Amazingly enough, a machine like the Large Hadron Collider in Switzerland can reproduce conditions of the early universe when it was 10^{-10} seconds old. For a brief moment, those conditions exist in a volume the size of a proton deep inside the machine. Unfortunately, at the moment we don't yet know how to build machines with enough oomph to replicate conditions any closer to the Big Bang itself. And so the unification of the electromagnetic and

ALBERT EINSTEIN'S FAILED QUEST

Albert Einstein spent much of his later years trying unsuccessfully to produce a unified field theory. But how did a physicist of his caliber fail at this task? It turns out that Einstein was trying to unify the wrong forces. He concentrated on how gravity and electromagnetism might marry, a question we have yet to answer. In Einstein's defense, at the time, the weak and strong forces had only recently been discovered and were yet to be deeply understood.

weak forces—10^{-10} seconds after the Big Bang—marks the point beyond which we cannot test our ideas through experiments.

THE UNIFICATION OF THE STRONG FORCE

Resuming our journey backward, we must cross a huge interval of time before we find another crucial change—all the way back to 10^{-35} seconds after the Big Bang. This number represents a period of time so tiny that there is nothing in our human experience with which to compare it. Nevertheless, our best theories tell us that this was when the basic bricks of our universe were formed: the time when the six quarks and six leptons first showed up.

Of course, whatever our ideas tell us about the behavior of the universe in this early epoch, we have no way to check those hypotheses experimentally. We do, however, have a theory

Photographed together in 1930, Niels Bohr (left) and Max Planck had by that time both received Nobel Prizes for their contributions to the development of quantum mechanics.

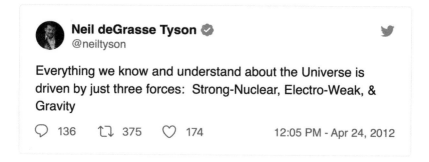

Neil deGrasse Tyson
@neiltyson

Everything we know and understand about the Universe is driven by just three forces: Strong-Nuclear, Electro-Weak, & Gravity

136   375   174   12:05 PM - Apr 24, 2012

called the standard model of particle physics, which nicely explains the behavior of matter at energies accessible to modern accelerators (and then some). But even optimistic physicists will not push the theory earlier than 10^{-35} seconds after the Big Bang. So for now, we think of this time as the limit of any understanding that carries confidence.

That moment—10^{-35} seconds after the Big Bang—represents a turning point. Before then, only two forces acted in the universe: gravity and the strong force unified with the electroweak. Thus, events at this time can be thought of as another example of simplification, a moment after which many more complex features emerged that would become fixtures in the modern universe.

Theoretical physicists are nothing if not inventive, however, and the fact that we have no good ideas that take us farther back in time doesn't mean we have no ideas at all. In fact, many have been proposed. We'll look at some of the more interesting hypotheses in the next chapter—but note that most of the community of physicists expects the next and final unification of forces to occur at 10^{-43} seconds. We call this the Planck time, to honor the German physicist Max Planck, who is generally recognized as the founder of quantum mechanics. For all times earlier, we expect that gravity and the strong-electroweak force were unified as one. The large-scale universe, nicely explained by general relativity's description of gravity (the theory of the large), now occupies tiny volumes controlled by quantum mechanics (the theory of the small). To understand this shotgun marriage, we desperately need a quantum theory of gravity.

Before we take that final step, however, let's raise one troubling fact: that matter exists in our universe at all.

THE ANTIMATTER PROBLEM

Why does matter exist at all?

Our entire understanding of mass and energy in the universe tells us, guided by $E = mc^2$, that when high-energy light (x-rays and gamma rays) becomes mass, it spontaneously creates pairs of matter and antimatter particles. Thereafter, when either particle finds its matched antiparticle, they annihilate, becoming pure energy once again.

When the early universe was hot enough for x-rays and gamma rays to dominate the energy spectrum, this two-way street—energy becoming matter, becoming energy, becoming matter—unfurled continually. But as the universe expanded, it cooled to energies that could no longer spontaneously create matter–antimatter particle pairs. There are no known particles with mass that low.

That's odd, when you think about it. If all mass created in the universe comes from matter–antimatter pairs spawned from x-rays and gamma rays, then what actually happened when the universe cooled so much that there were no x-rays or gamma rays left? Under that scenario, every particle ever created has an antimatter counterpart. As the universe continues to cool, particles eventually meet up with their antiparticles and annihilate for one last time, leaving a universe of light—and nothing else.

So modern physics tells us the universe should have no matter of any kind in it at all. But we do. In fact, we're all made of matter, and not antimatter. So somehow, at some point in

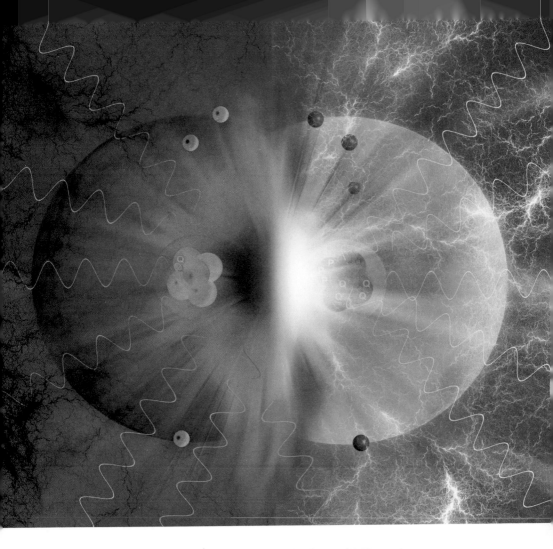

Matter and antimatter mirror each other in this illustration,
with electrons, quarks, and other subatomic particles on one side,
their antiparticles on the other.

the early universe, a few x-rays or gamma rays spontaneously
made just one lone matter particle, causing a profound asym-
metry to the balance of particles in the universe. To account
for what we see, calculations require that about one in a hundred million of these conversions creates a lone matter particle,
violating key conservation laws of particle physics. Perhaps
this was another one of those mysterious but real changes in
the early universe.

That's the explanation we're going with. And we're sticking to it, for now.

THE GRAND SCENARIO

We can finally piece together the entire history of the universe as we understand it, down to when our theories start to break down. Let's replay the story now.

10^{-43} SECONDS AFTER | Everything we describe up to this moment—the Planck time—is speculation. We have no experimental data or solid theories to guide us. Nevertheless, we expect that only one force existed in the universe. Some of our theories also predict that there was only one kind of particle as well. If this is true, then the universe started out in the simplest state imaginable: a universe composed of one kind of particle interacting through one kind of force.

10^{-35} SECONDS AFTER | Here the universe undergoes a slew of changes. For one thing, it rapidly expands during a time called inflation. Slightly more matter than antimatter is produced—specifically, 100,000,001 matter particles for every 100,000,000 antimatter particles. The antimatter–matter particle pairs find each other and annihilate to produce intense radiation, which eventually becomes the cosmic microwave background. The strong force splits away, leaving the universe with three operating forces instead of two: gravity, strong, and electroweak.

10^{-10} SECONDS AFTER | Here the electromagnetic force splits away from the weak force, marking the final splitting of the forces from three into the four we see today: strong, weak, gravity, and electromagnetic.

10^{-5} SECONDS AFTER | This time marks the arrival of familiar particles and atomic structures. Quarks assemble to form the elementary particles found within the nucleus, such as protons and neutrons. Once inside the elementary particles, the quarks are confined and no longer exist on their own.

THREE MINUTES AFTER | Here the temperature has dropped enough to enable a proton and a neutron to combine and form a simple nucleus, without subsequent collisions knocking them apart. For a brief period, nuclei of atoms up to lithium in complexity are formed. The Hubble expansion then spreads particles too far apart for them to continue building nuclei. After three minutes, matter and radiation coexist in the form of a plasma, and any attempt for matter to clump is swiftly blown apart by the high-energy radiation. Dark matter, on the other hand, gathers under the influence of gravity, invisible and impervious to the destructive powers of radiation.

380,000 YEARS AFTER | Atoms form as electrons attach to nuclei, freeing the light and rendering the universe transparent. This radiation manifests today as the cosmic microwave background, and ordinary matter clumps to form stars and galaxies wherever dark matter is found.

In short, everything began in the simplest possible state and, from there, expanded and evolved into the complex universe we know today.

THE END OF KNOWLEDGE

Well, that's it. We've come as far in understanding the origins of the universe as we can. Beyond this point in our journey, only hypotheses and conjectures remain—nothing more. And, as any curious person knows, the most interesting questions are the ones we don't yet know to ask.

Standing at our milepost of 10^{-35} seconds and looking backward in time, we can see two mountains we'll have to climb. The first is what does or doesn't happen at and before 10^{-43} seconds, the Planck time. As previously noted, the consensus among theoretical physicists is that the trend line of unified forces will continue, and that at the Planck time, the gravitational force will unify with the strong-electroweak.

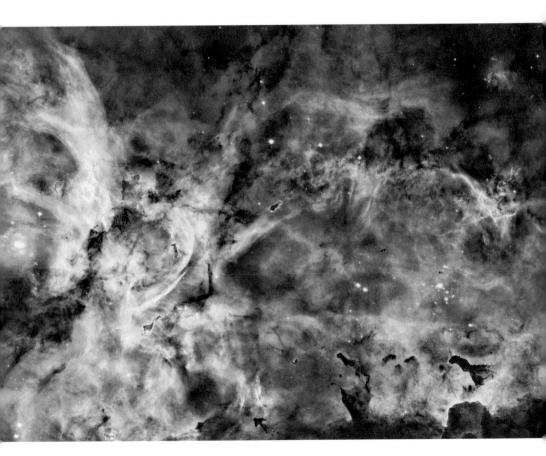

A Hubble Space Telescope mosaic shows the Carina Nebula, a stellar nursery whose massive stars are shredding the very gas cloud that gave birth to them.

We know that at the quantum level, the exchange of virtual particles generates the strong, electromagnetic, and weak forces. According to general relativity, however, gravity results from the warping of space-time by the presence of matter—a force at work with nary a mention of virtual particle exchanges. Thus, creating a theory of gravity that works in the quantum world isn't just a matter of fiddling around with equations; it involves reconciling two fundamentally different ideas of how forces can be generated.

But even if we solve this problem, we are left with a higher mountain to climb: the nature of the initial event itself. If we think that time began at the Big Bang, then the question "What preceded the Big Bang?" doesn't make sense. Additionally, according to some ideas, the exact nature of the universe isn't locked into one set of physical laws but can vary. In this situation we can have many coexisting universes: a multiverse. These are universes that exist side by side, each governed by its own slate of cosmic laws, with no two of them ever meeting.

Fasten your seat belts, then, for a journey into unknown unknowns.

HOW WILL IT

ALL END?

9

When we think about the end of the universe, we naturally first wonder about our own solar neighborhood—and the end of the Sun. Like all stars, our Sun deploys one tactic after another to hold itself up against the force of gravity. Each new strategy shapes an episode in the saga of its death.

As previously explained, the solar system began as a collapsing interstellar cloud, governed by the inexorable inward pull of gravity. When the resulting compression raised the core temperature high enough, nuclear reactions ignited, fusing hydrogen into helium. Specifically, four protons combined and transformed to become a single helium nucleus containing two protons and two neutrons, plus a spray of other particles, plus energy.

The resulting outpouring of radiation balanced the inward pull of gravity and the Sun became stable. The Sun has been fusing hydrogen in its core for about 4.5 billion years. In about five billion more years, it will have consumed all the hydrogen

Some five billion years hence, the Sun will expand into a red giant, engulfing the inner planets. This futuristic view from Earth's parched surface shows the Moon as a black disk against the Sun's roiling surface.

in its core, and the force of gravity—kept in check for almost 10 billion years—will reassert itself, threatening collapse.

There are, in fact, two sources of energy available to counteract this collapse. One is the pristine, unfused hydrogen in a shell around the core. The other is a series of nuclear reactions that convert the helium ash to carbon in the core. For that, the proton inventory checks out: Three helium nuclei, each with two protons, fuse to become one carbon nucleus containing six protons.

The Sun employs both these tactics as it enters old age, resulting in a complex series of events. The most important of these are (1) the Sun will lose about a third of its mass from a vastly increased solar wind, and (2) the outer envelope of the Sun will expand and cool as it transforms into a new category of star called a red giant, sequentially engulfing the orbits of Mercury and Venus and threatening to do the same of Earth. Eventually all this expanding gas escapes into interstellar space, laying bare a tiny, stable, hot stellar corpse.

Best to be somewhere else when all this unfolds.

A star like the Sun isn't massive enough to ignite more nuclear reactions beyond converting helium to carbon, so there is nothing to prevent the inevitable collapse. At this point, however, a new character enters our saga. Electrons that were torn from their atoms early on and have been in the plasma background for billions of years now step up.

The laws of quantum mechanics tell us that a cloud of electrons cannot be compressed indefinitely. You can say that each

HOW OLD ARE WE?

If the lifetime of the universe were compressed into a single year, the Sun and the solar system were born in early September—fairly late in the lifetime of the Milky Way.

electron requires some elbow room to maintain its identity. The force of gravity can compress what remains of the Sun to about the size of Earth before running into the outward pressure of these electrons—and the compression stops forever. The Sun terminally becomes a star called a white dwarf, an ember slowly cooling in the sky. And the saga of its life ends here.

THE END OF EARTH

So while the Sun battles for life against its own collapse, what's happening on our home world?

The most striking change comes from a rise in the Sun's luminosity. Although the Sun has been generating energy by fusing hydrogen in its core for 4.5 billion years and will continue to do so for another five billion, it grows brighter all the while. When the planets formed, for example, the Sun was about 30 percent fainter than it is today and has steadily brightened, becoming about two-thirds again more luminous by the time the hydrogen in its core is used up. So how will the planet respond to this kind of warming? To be clear, life as we know it would not survive this new threat—at least not on Earth. And no, the warming we are experiencing today cannot be explained by the Sun's stately life cycle.

Ignoring whatever humans might do to Earth over the next few million years, our planet won't look much different than it does today. Ice ages, which depend primarily on the details of Earth's rotation and orbit around the Sun, will continue to come and go. They will, however, become less frequent as the Sun warms the planet. The slow, steady motion of the continents, driven by convection in Earth's mantle, will be unaffected in the short term. Geophysicists have suggested, in fact, that in another 250 million years, all the continents will be reunited into one supercontinent: a reprise of what existed 250 million years ago in the form of Pangaea.

> **Neil deGrasse Tyson** ✓
> @neiltyson
>
> In five-billion years, as the Sun begins to die, its outer layers of glowing plasma will expand stupendously, engulfing the orbits of Mercury, then Venus, as the charred ember that was once the oasis of life called Earth vaporizes into the vacuum of space.
>   Have a nice day!
>
> ◯ 7.2K   ⭗ 55.7K   ♡ 204K   7:56 AM - Mar 12, 2018

A billion years from now, the planet's average temperature will have risen above our own body temperature. At those levels, evaporation rates increase, and more moisture enters the atmosphere from our oceans, seas, and lakes. Ultraviolet radiation from the Sun will then break apart those H_2O water molecules into hydrogen and oxygen atoms. The lighter, faster-moving

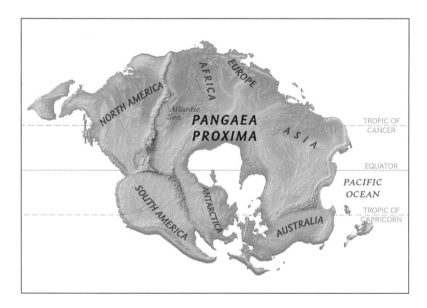

The continents are ever moving on Earth's surface. Geologists predict they will ultimately reconnect into a single landmass, dubbed Pangaea Proxima.

hydrogen atoms will escape to space as the oceans disappear entirely, desiccating the entire planet. Volcanoes will continue to erupt, spewing deep stores of water and carbon dioxide into the atmosphere. But without the oceans to help uptake atmospheric carbon dioxide, as they do continuously today, a strong greenhouse effect will ensue. With ultraviolet light from the Sun continuously killing the H_2O molecules, the sub-crystal lubrication provided by water will cease, forcing plate tectonics to grind to a halt as the continents lock into place. Eventually, about three to four billion years in the future, the runaway greenhouse effect will raise the surface temperature of the planet high enough to liquefy our rocky surface into a lava ocean.

The final act will come when the Sun enters its red giant phase and its outer regions swell to extend beyond Earth's orbit. Because at that point the Sun will have lost mass from increased solar wind, its gravitational grip on the planets will be reduced, sending Earth into a steadily larger orbit. If the Sun does not actually swallow it up altogether, Earth will be fated to circle the white dwarf as a dead, charred cinder.

This might be a good time to ask, "How's that space program coming along?"

PANGAEA

Some 250 million years ago, all the land on Earth was connected into a supercontinent called Pangaea, ancient Greek for "all-Earth." Then magma welled up from beneath the crust, creating rifts and diverting landmasses into the seven continents we see today. Evidence of identical sedimentary layers and fossil species on the coastlines of faraway continents affirm this phenomenon.

In another 250 million years, geologists predict, the continents will re-form into a single supercontinent, resulting in a hypothetical landmass dubbed Pangaea Proxima, the next Pangaea.

UNPREDICTABLE DOOMSDAY: VOLCANOES

The Sun's evolution isn't the only threat to the welfare of life as we know it. Earth abounds with perils lying in wait—like volcanoes.

We live on a planet that was once entirely molten and is still cooling off. Liquefied rock, called magma, transports heat to the planet's surface, producing all sorts of phenomena, from hot springs to volcanic eruptions. Most of the time these eruptions, no matter how spectacular, affect only a relatively small area around the volcano. Occasionally, however, volcanic eruptions have global consequences. In 1815, for example, Mount Tambora, in what is now Indonesia, erupted and produced enough particles in the stratosphere to block incoming sunlight. The entire Earth cooled and turned 1816 into what came to be called the year without a summer.

Occasionally, however, the magma coming up from the interior cannot punch out from under Earth's crust to make an isolated volcano. In this case, the pressure grows until large areas of Earth's surface rupture in a titanic explosion and a massive outflow of lava. If the volume of that outflow is greater than about 1,000 cubic kilometers—you read that correctly: 1,000 cubic kilometers—the event is called a supervolcano, and it can spew enough material to bury an area the size of Texas under five feet of rock. There are 20 active supervolcanoes in the world. The most familiar, perhaps, is in Yellowstone National Park in the United States. This supervolcano last erupted 664,000 years ago. If a similar eruption occurred today, much of North America would be buried under volcanic ash.

Geologists have found evidence for no fewer than 47 eruptions of supervolcanoes in Earth's past—most recently, 26,500 years ago, in New Zealand, around the age of human cave dwellers. So despite what you may see in apocalyptic sci-fi films, life on Earth (including the humans) will most likely survive the

Neil deGrasse Tyson ✓
@neiltyson

The stark beauty of Yellowstone Park, atop a dormant super-volcano. Reminders that not all of our planet is a haven for life. Across many parts of Earth's surface, "Mother Nature" will just as soon see you dead.

○ 432   ⟲ 1.7K   ♡ 12.7K   5:35 PM - Oct 21, 2018

next eruption, even if not all organisms do (especially those who decide to get a closer look).

Millions of years ago, yet another scale of volcanic activity known as large igneous provinces once threatened Earth's inhabitants. These eruptions dwarf supervolcanoes and can release hundreds of thousands of cubic miles of lava. The Deccan Traps, for example—an extensive geological formation in west-central

India—date back to eruptions that occurred about 65 million years ago, while the Siberian Traps—a similar geological formation in northern Russia—originated from eruptions 250 million years ago. The timing of these outflows matches the two biggest mass extinctions in the history of life on Earth, and they cannot be ignored when looking for a smoking gun to answer why these extinctions occurred.

Yellowstone National Park sits atop a supervolcano—a massive cauldron of magma that last erupted more than half a million years ago. Today, the supervolcano forms a steamy, sulfurous landscape wrought with underground quakes.

THE CATACLYSM OF MOUNT KRAKATOA

The red sky in Edvard Munch's painting "The Scream" is thought to be a result of material thrown into the atmosphere by the explosion of Mount Krakatoa in Indonesia in 1883. When it erupted, Krakatoa was actually torn open. Seawater rushed into the gap, causing an explosion massive enough to be heard in Australia, 2,000 miles away—the loudest sound in recorded history.

UNPREDICTABLE DOOMSDAY: IMPACT

On an ordinary day 50,000 years ago, an object the size of a 16-story building fell out of the sky over what is now the U.S. state of Arizona. Composed mostly of sturdy iron, the asteroid survived its fiery descent through the atmosphere and vaporized on impact, blasting out a mile-wide crater—and creating a popular tourist attraction. The Barringer Crater, named for its landowner at a time when the crater's origin was a bit of a mystery, is now more appropriately called Meteor Crater. It serves as a poignant reminder that Earth inhabits a very dangerous region of space.

Most asteroids in the solar system orbit between Mars and Jupiter, a region named the asteroid belt. (As with Meteor Crater, in astrophysics we tend to be sensible about the names we give things.) Occasionally, however, a random collision or the gravitational influence of a planet kicks an asteroid toward Earth. These asteroids aren't aiming for us, but if we happen to get in their way—if we happen to be in the wrong place at the wrong time—they will hit us, just like that hunk of iron hit Arizona.

The most catastrophic impact we know of occurred 65 million years ago, when an asteroid about the size of Mount Everest hit Earth near a place now called Chicxulub on Mexico's Yucatan Peninsula. An asteroid this size, moving upwards of 10 miles a second, packs a punch with a thousand times more energy than is contained in the entire human nuclear arsenal.

> **Neil deGrasse Tyson** ✓
> @neiltyson
>
> Arizona is famous for its holes in the ground. Grand Canyon took millions of years to form. Meteor Crater took a few seconds.
>
> 💬 119   🔁 1.2K   ♡ 2.7K   8:06 PM - Feb 1, 2015

This kinetic energy was converted to heat on impact, blowing out a crater more than a hundred miles across and triggering the most recent of five mass extinction events. Dust from the explosion blanketed the upper atmosphere, blocking sunlight for years. The darkness, together with the tsunamis and fires ignited by falling debris, wiped out two-thirds of all the species of life on the planet, including all the classic dinosaurs we've come to love (and fear) from childhood.

So let's ask the unnerving question—is there an asteroid out there with our name on it? Does a so-called extinction-level event await?

NASA has dedicated several programs to detecting Earth-threatening objects. The best known is the Panoramic Survey Telescope and Rapid Response System (Pan-STARRS), which consists of telescope facilities located near the summit of Hale-akala in Hawaii. Already, the project and others like it have discovered hundreds of thousands of asteroids, tens of thousands of which are classified as risky Near-Earth Objects, or NEOs. We expect this to be a fairly complete catalog of all the potential threats greater than the size of a football stadium.

If we do find such a threat, we probably won't try to blow it out of the sky, Hollywood style. Nukes would cause a chaotic

Arizona's Meteor Crater shows the damage caused by an asteroid that struck Earth 50,000 years ago.

and incalculable ricochet of debris—potentially launching deadly chunks of the thing even faster toward us. More likely, we'll find ways to nudge the asteroid, bit by bit, until it veers away from a collision course with Earth.

THE COMING COLLISION

In spite of the cosmic expansion, discovered by Edwin Hubble, in which galaxies all recede from one another, galaxies that happen to be close to one another will feel a mutual gravitational attraction that is strong enough to override the local cosmic expansion. In other words, galaxies collide. In fact, astrophysicists have suspected for decades that the Milky Way and our twin neighbor, the Andromeda galaxy, will collide in a few billion years.

The details of this collision have become clearer in the last few years, thanks to new data gathered by the Gaia spacecraft, which we met in an earlier chapter during our discussion of parallax and the distance ladder. Launched by the European Space Agency in 2013, Gaia's mission is to assemble a three-dimensional map of our galaxy by measuring the position in space of a billion stars with unprecedented accuracy. And although the primary function of the spacecraft is to look at stars in the Milky Way, Gaia can also detect light from bright stars in Andromeda. Based on these measurements, we can predict a scenario of the coming collision.

About 4.5 billion years from now, the galaxies will approach each other for a sideswiping encounter. If we look only at the luminous parts, we'd call it a near miss. But a massive halo of dark matter surrounds each galaxy, and that's the gravity that matters here. After the Milky Way and Andromeda pass by each other, the attraction of those halos will cause us both to slow down, stop, and reverse course, colliding once again.

HOW EMPTY ARE GALAXIES?

If the entire continental United States of America contained just 30 bumblebees, any two of them would have a far better chance of randomly bumping into each other than would two stars in colliding galaxies. Can't picture that? How about this: If the Sun were the size of the period at the end of this sentence, then its nearest star would be four miles away.

Note that galaxies are not solid objects. In fact, there's mostly empty space between their stars. So a collision is not a onetime crash. The galaxies will emerge and get pulled back again, repeating in ever smaller recoils, until the system settles down into a single jumbo galaxy—named, playfully if unimaginatively, Milkomeda.

OPEN, CLOSED, OR FLAT?

When thinking about the fate of the entire universe, rather than the fate of our own neighborhood, we must first understand the universe's basic geometrical structure. Imagine throwing a ball up from Earth's surface. Unless you're Superman or Captain Marvel, it eventually falls back to Earth as gravity slows it down. Throw the ball upward with that same speed but from the surface of a small asteroid, and it will fly off into space, never to return. If you throw the ball at a perfect velocity to complement and counteract the pull of gravity, the ball might get stuck in an orbit. Thus, the fate of the ball depends on the speed and direction at which you launched it and the gravitational force it feels along the way.

We can make an analogous argument about the fate of the Hubble expansion of the universe. If the universe has enough mass—enough gravity—to slow down those outrushing galaxies and turn them around, the expansion will stop and be reversed someday. We call this a closed universe. If, on the other hand,

80 TO 100 TONS

The amount of space material that falls through Earth's atmosphere every day

the universe doesn't have enough mass, the expansion will continue forever. We call this an open universe. The transition point between these two cases is called a flat universe—where it has precisely enough mass to stop the expansion and eventually stay in balance between the two scenarios. The amount of mass to accomplish this is said to "close" the universe.

So, when we think about the future of the universe we have to consider three things: (1) the amount of ordinary matter, (2) the amount of dark matter, and (3) the amount of dark energy.

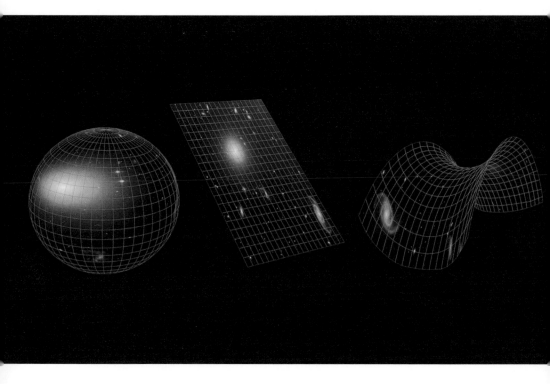

The shape of the cosmos influences how it will end. Closed (left), flat (center), or open (right)? All evidence points to a flat universe.

The first two of these act through the traditional attractive gravitational force and, together, slow down the expansion. Dark energy, however, is a kind of antigravity operating in the vacuum of space, acting to speed up the expansion.

There is another way to think about the long-term future of the universe, however—a way that depends on geometry rather than the actions of gravity. Here, opposite, are three possible shapes for the basic structure of the universe: a closed universe, a flat universe, and an open universe.

So, which shape represents the actual universe we live in? We ask not a theoretical question but an empirical one. We need to measure something that allows us to choose among these scenarios.

Referring to high school geometry, we could see if two parallel lines remain parallel to each other over a distance. From the figure opposite, it's clear this can only happen in a flat geometry. A closed geometry would see the lines swing back to meet each other, as do lines of longitude on Earth's surface. And although it's not technically possible to do such a measurement by sending laser beams out and seeing how they behave over long distances, we have at our disposal photons that have been traveling for more than 13 billion years—our friend, the cosmic microwave background. In fact, astrophysicists analyzing those microwaves have resolved the question: We live in a flat universe.

This simple fact deeply informs our sense of how the universe will end.

WHAT IS THE MATTER?

Once we know that our universe is flat, and that there is not enough gravity to close it, the next question that comes to mind is: How big a slice of cosmic pie are we? Let's look, in turn, at each slice of what composes the universe.

BARYONIC MATTER: 5 PERCENT ‖ Baryonic matter is stuff made of ordinary particles, the kind of matter we're used to. Stars, planets, asteroids, comets, interstellar dust clouds, black holes, and human beings are all made of baryonic matter.

This familiar form of matter makes up only about 5 percent of the universe—a sobering fact. That means the entire history of science, from the earliest civilizations to the present, has been devoted to exploring only a tiny fraction of all that exists. On the other hand, we have—over the millennia—acquired a pretty good idea of what baryonic matter is and how it behaves. We're quite proud of what we can do with that 5 percent.

DARK MATTER: 27 PERCENT ‖ You may remember, this really should have been called dark gravity, because that's precisely what it is. But we're stuck with the term "dark matter," so we must roll with it, constituting a bit less than 30 percent of the universe.

Even though we don't know what dark matter is, we know what it does—and, more important, what it doesn't do. We know it exerts a gravitational force, and we can see the effects of this force in things like the rotation of galaxies, the structure of galaxy clusters, and the bending of light rays in a phenomenon known as gravitational lensing, first predicted by Einstein.

Dark matter does not interact with electromagnetic radiation (light), and theorists have several notions of what it might be. Most current ideas assume that dark matter is composed of some

DEAP-3600, an exquisitely sensitive dark matter detector, operates more than a mile underground in a nickel mine in Ontario, Canada.

sort of elusive elementary particles, but sophisticated experiments have yet to detect them.

DARK ENERGY: 68 PERCENT | Dark energy constitutes both the biggest slice of the cosmic pie and the slice we know the least about. We know that it acts as a kind of antigravity, pushing galaxies apart, accelerating the Hubble expansion—but that's about it. At the moment, two leading theoretical contenders vie for the identity of dark energy.

The most popular is that it represents the energy of empty space. In an early version of general relativity, Einstein included a term called the cosmological constant to allow for this possibility. Physicists who have tried to use quantum mechanics to calculate this energy find their answers are off by a factor of 10^{120}—that's a one followed by 120 zeroes. It's hard to get a mismatch that large between theory and observation. In fact, it's the most wrong calculation there ever was—ever.

An alternative candidate for dark energy, called quintessence, borrows its name from ancient Greek philosophy, which held that in addition to the four familiar elements of earth, fire, air, and water, a fifth element existed only in the realm of the gods. Theorists suggest that dark energy might be a new kind of space-filling fluid whose effect is to speed the expansion of the universe.

Whatever dark energy is, it makes up about two-thirds of all that drives the universe.

EINSTEIN'S BLUNDER

Einstein introduced the cosmological constant because he presumed, like everyone else, that the universe was static and stable, but to maintain those states, he needed a force to counteract gravity. Had he not been swayed by this bias, he could have predicted that the universe was either expanding or contracting. He swiftly dropped the concept when Hubble discovered the universal expansion and later called his cosmological constant "the greatest blunder of my life."

But because a cosmological constant does indeed exist, serving not to balance cosmic gravity but to override its influence altogether, Einstein's greatest blunder may have been saying that the cosmological constant was his greatest blunder. In other words, even when Einstein was wrong, he was right.

OPTIONS FOR THE END

Traditionally, stories about the end of the universe begin with possibilities corresponding to open, closed, and flat systems. In a closed universe, the Hubble expansion stops and reverses itself at some future date. Matter rushes inward, collapsing to that original unfathomably dense beginning. This scenario is called the Big Crunch, which may be followed by a new expansion, the Big Bounce. So a closed universe would be the precursor to a cyclic universe. Intriguing, but all data ever collected tell us that we don't live in a closed universe.

The remaining two possibilities for the end of the universe, given that it is either flat or open, depend on the nature of dark energy. Early in its history, the universe had a lot of ordinary and dark matter jammed together, so that gravity was the dominant force in any volume of space, overcoming any manifestation of dark energy at the time. Consequently, the universe put on the brakes, slowing the Hubble expansion. By the time the universe was about five billion years old, the ordinary and dark matter had spread thin, weakening the dominance of gravity over all things. This paved the way for dark energy to dominate gravity, forcing an increase in the expansion rate. We're still in this era of accelerated expansion governed by dark energy.

Here, we wonder whether or not dark energy is finite. The fate of the universe depends on the answer.

If the amount of dark energy is finite, then its effects will dilute as space expands. In this case, gravity will eventually take over again and the expansion will slow down but will never end, as predicted for a flat universe.

Perhaps, however, the amount of dark matter increases as the amount of space grows—maybe it's a property of the vacuum itself. The more the universe expands, the thinner gravity gets but the more vacuum there is, increasing the strength of dark

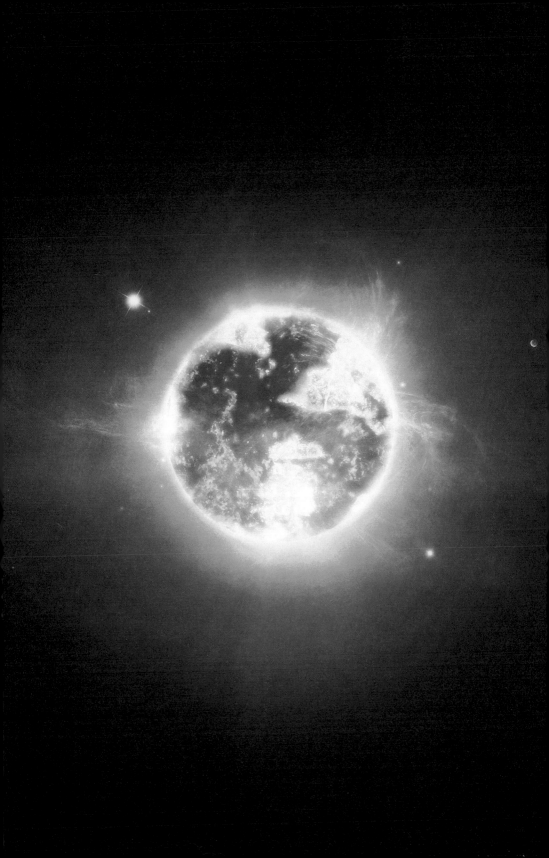

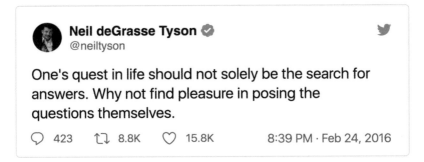

One's quest in life should not solely be the search for answers. Why not find pleasure in posing the questions themselves.

423   8.8K   15.8K   8:39 PM · Feb 24, 2016

energy relative to gravity. This expansion will keep accelerating in a runaway scenario, leading to a scary fate called the Big Rip, which some theories predict will begin in about 22 billion years.

In this development, the separation between galaxies will increase at first, but then the expansion of space will start to separate the stars within galaxies. Three months before the Rip, solar systems dismantle. Thirty minutes before the Rip, the space between atoms in materials will increase to the point that all material objects—planets, rocks, people—will tear apart. In the end, the relentless force of antigravity will even tear those atoms apart, leaving an ever emptying collection of elementary particles in its wake.

Until we know a whole lot more about dark energy than we do now, we won't be able to choose between the various end states of the Hubble expansion.

THE EDGE OF THE MAP OF TIME

We have reached the limits of our theoretical knowledge. Beyond this, we're compelled to emulate the technique of medieval mapmakers who, when they reached the end of known territories, would write on the edge of the map, "Here be dragons," and go home.

Our Sun, just one star amid a billion trillion in the universe, will expand in death into a red giant, eventually laying bare its fuel-exhausted core, called a white dwarf.

WATCHING THE END OF THE UNIVERSE

The lamps are going out all over Europe.
—BRITISH DIPLOMAT SIR EDWARD GREY, 1914

It's one thing to talk about the end of the universe in abstract theoretical terms, but quite another to imagine what it might look like to an observer on Earth. For the sake of argument, let's pretend that this observer is suitably shielded from the laws of physics, which are bringing an end to everything else, and blessed with a sufficient life span to watch the entire show. Although what we see in the sky may be punctuated with various kinds of fireworks, the overall effect will be for the skies to gradually darken. Sir Edward Grey's quote, repeated above and uttered at the beginning of World War I, captures the desolation of the future universe.

The first few billion years will be unexceptional as the Sun gradually brightens and Earth feels the concomitant effects. But thereafter, we will begin to notice something new in the sky. As the Andromeda galaxy approaches in preparation for its collision with the Milky Way, it will become a larger and larger fuzzy patch of light. As the collision approaches, Andromeda's shape, as well as that of the Milky Way, will begin to distort in the sky, influenced by our mutual gravitational attraction. Oddly enough, as we have pointed out, the serial collisions of the galaxies probably won't have much of an effect on the Sun or our solar system; stars are just too widely spread for that to matter.

When the Sun enters its final stages as a red giant and then a white dwarf, we'll notice something else: The stars begin to disappear.

Every star begins life by fusing hydrogen and ends life as a white dwarf or supernova or black hole when its nuclear fuels run out. Five billion years from now, our Sun will become a cooling cinder in space, supported against gravity by its electrons. The

THE FAR DISTANT FUTURE

We can make some predictions about the next few billion years based on solid physics. Beyond that, our ideas depend more and more on conjectures about the nature of dark energy and elementary particles. With that disclaimer, we present a timetable for the far and distant future, estimated in years from the present:

- **1 billion**—Earth loses its oceans.
- **4.5 billion**—Sun reaches peak of red giant phase. Mercury, Venus, and Earth are swallowed.
- **5 billion**—Andromeda galaxy collides with the Milky Way.
- **6 billion**—Sun becomes a white dwarf.
- **22 billion**—Start of the Big Rip. The end.

If no Big Rip . . .

- **100 billion to 150 billion**—All galaxies beyond Local Group leave the observable universe.
- **450 billion**—All galaxies in the Local Group coalesce into a single galaxy.
- **100 billion to 1 trillion**—The last waves of star formation that the universe will ever see.
- **1 trillion**—The longest-lived stars in the universe begin to die. All stars that can be made have been made, as the universe plunges into darkness.

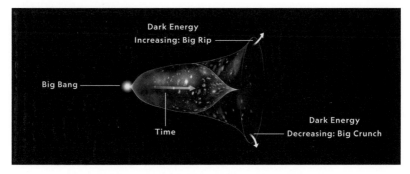

A graphic representation of all time, depending on which way things go—from the Big Bang (at left) to either the Big Rip, arching outward, if dark energy increases, or the Big Crunch, funneling inward, if dark energy decreases

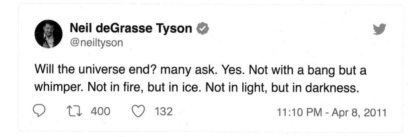

Neil deGrasse Tyson ✔
@neiltyson

Will the universe end? many ask. Yes. Not with a bang but a
whimper. Not in fire, but in ice. Not in light, but in darkness.

💬 	🔁 400 	♡ 132 	11:10 PM - Apr 8, 2011

electrons won't quit supporting it, but eventually the Sun will cool
to the temperature of the universe and stop radiating. It goes dark.
The same fate awaits any star, no matter what path it follows to
the end.

Furthermore, as the stars die, we notice that the distant galaxies are disappearing as well. The effect of dark energy increases
the space between galaxies in an overall acceleration of the
Hubble expansion. Eventually, each galaxy will reach a point
where the expansion of space between it and Earth becomes so
large that the light it emits will never reach us. Like the stars,
the galaxies wink out of existence, one by one.

In the end, a cold, dark universe surrounds us, populated by
a thin soup of elementary particles and decaying black holes.

Nighttime northern lights gently loom over ice cracks in Alberta's Abraham
Lake, evoking the way the universe will end: in ice and darkness.

WHAT DOES NO
DO WITH EVERY

The Jewel Box, a loose cluster of stars in the constellation Crux, the Southern Cross

THING HAVE TO
THING?

10

By now, we're heavily invested in understanding what the universe is made of and how it works. But like the black type on this page, you see letters and words only where there's an absence of light. We call it ink. But optically, it's all places where the white of the page has been blocked from reaching your eyes. In books, the nothing is everything. So, too, in cosmology: It's hard, if not impossible, to talk about the existence of anything without tandem references to the existence of nothing. Like yin and yang, they complete each other.

Aristotle tried to understand nothing, as manifested by the absence of air. He confidently proclaimed that "Nature abhors a vacuum." His argument was simple: If a vacuum was created, then the surrounding air would rush in and obliterate it, proving nature's abhorrence.

During the Middle Ages, this argument acquired a theological overlay when a vacuum was taken to symbolize the absence of God. The medieval Latin name for a vacuum—*horror vacui*—requires no translation. In fact, in the Condemnation of 1277,

With all we know representing only about 5 percent of the universe—
the colored jelly beans in this jar—what is the rest?

Stephen Tempier, the bishop of Paris, named the belief in the existence of the vacuum as one of 219 errors condemned by the Church—right up there with fortune-telling and incantations that summon the devil.

We can point to the German diplomat-scientist Otto von Guericke as the man who demonstrated the reality of the vacuum. In 1654, he performed a famous experiment where he simply put two metal hemispheres together, pumped the air out from between them, allowing atmospheric pressure surrounding the sphere to do its thing, and showed that teams of horses (30 in all) couldn't pull them apart. With this demonstration, the notion of nothing changed, and the vacuum version of nothing became a standard tool in experimental science.

The largest vacuum system in the world today is along the ring of the Large Hadron Collider in Switzerland. It's actually three separate systems—two to provide thermal insulation, like a thermos, and one to clear atmospheric molecules and any other stray atoms from the beam ring, so particles can pass unimpeded. It takes a full two weeks to pump the air out of the vacuum system—but when that is done, the residual pressure in the beam ring is 10^{-13} atmospheres. For every 10 trillion molecules, you'll now find only one.

In fact, the space inside the beam pipes in the LHC is emptier than interplanetary space, making it the emptiest place in the solar system. Clear sailing for the high-energy particle beams. You don't want them slamming into stuff they're not supposed to.

NOTHING AIN'T WHAT IT USED TO BE

The development of quantum mechanics in the 1920s uprooted many scientific concepts of the day—and the concept of the vacuum was not spared.

From Aristotle up to that point, whether or not you believed

Neil deGrasse Tyson ✓
@neiltyson

"Nature abhors a vacuum" came from space-illiterate people. In fact, Nature just loves a vacuum. It's most of the universe.

💬 339 🔁 1K ♡ 1.8K   11:54 AM - Jun 17, 2013

that the vacuum was a real thing, you thought about it in the same way. A vacuum is a region of space with nothing in it, period. That version of nothing was literally "no thing."

But the Heisenberg uncertainty principle changed all that. A particle can pop into existence out of nothing, so long as it disappears in a short enough time. We used the analogy of Cinderella at the ball—she could go, provided she got back by midnight. In the same way, we argued that virtual particles—particles that mediate the fundamental forces—are quantum fluctuations in the vacuum that can appear from nothing, provided they are reabsorbed within the time limits prescribed by the uncertainty principle.

The implications for the vacuum of this quantum concept are profound. The traditional vacuum is a static, lifeless thing. The quantum mechanical vacuum, on the other hand, is a dynamic place, teeming with particles that pop into and out of existence quickly enough to satisfy Dr. Heisenberg.

Air pump for vacuum experiments, circa 1800

Speaking of popping, imagine that you have a special kind of popcorn. The kernels in this popcorn not only pop as ordinary kernels of popcorn do, but they also un-pop—revert from the fluffy white treat to the original condensed kernel. What will you see if you put these imaginary kernels over a flame?

At first you'll see kernels popping up, pretty much at random. Soon, however, you'll see something else. One by one, again at random, the popped kernels will un-pop and return to their original state. Your spooky popcorn is like the quantum vacuum: On average, there is no gain or loss of energy in the system, and it's a far cry from the original Aristotelian nothingness.

This whole discussion sounds like something Alice would encounter in Wonderland—but rest assured, countless experiments verify the existence of this roiling quantum vacuum.

IS THE ENTIRE UNIVERSE A VACUUM FLUCTUATION?

American physicist Edward Tryon posed this question in 1973. He was the first person to investigate whether the laws of quantum mechanics might have something to do with the origin of the universe. He argued that there was no reason the universe couldn't have originated as a fluctuation—a rare one, to be sure—of the quantum vacuum.

This was an extraordinary suggestion. He was attributing the creation of the universe to the creation of a pair of virtual particles. The uncertainty principle, remember, sets a time limit on how long a virtual particle can last: The more massive a particle, the shorter-lived it will be. For an electron–positron pair, it's only about 10^{-21} seconds—a trillionth of a nanosecond. So how long could a universe possibly hang around whose mass equals a hundred billion galaxies? Surely not for billions of years.

More important, though, Tryon's argument seems to violate a fundamental law of nature: the conservation of energy. How could the mass of all those galaxies be conjured from nothing? You can trace this kind of objection at least as far back as the ancient Roman philosopher Lucretius, who famously argued *"Nil posse creari de nihilo*—Nothing can be created from nothing."

Neil deGrasse Tyson ✓
@neiltyson

If you seek only easy problems to solve, then ultimately, there'll be nothing about you to distinguish yourself from others.

💬 61   🔁 2.1K   ♡ 764   3:40 PM - Jun 27, 2012

The solution to this conundrum actually helps explain the origin of the universe.

When we look at the universe, we see two different kinds of energy. One, alluded to earlier, is the energy locked up in the mass of ordinary particles; this energy is positive. The other kind is the energy locked up in the gravitational fields; this energy is negative.

The ultimate in 21st-century vacuum experiments: One engineer bicycles past others inside the Large Hadron Collider, whose operations require a vacuum with pressure less than a billionth that of Earth's atmosphere.

THE USEFULNESS OF UNCERTAINTY

The Heisenberg uncertainty principle does not reveal a spooky, magic fact about the universe. Instead, it expresses a fundamental limitation on the act of measurement, which manifests most strongly in the world of small particles. The reason why you cannot know both the location and the speed of a particle at the same time is that the act of measuring one of them disrupts your ability to measure the other.

Think of retrieving a coin that slipped between the cushions of a car seat. As you reach for it, the width of your hand parts the cushions some more, allowing the coin to slip farther out of reach. The act of reaching for the coin made it harder to reach the coin.

Something else quantum physics has taught us is that you cannot know what you cannot measure. So Heisenberg's insight was to (successfully) elevate these practical facts to a principle of the universe.

Negative energy?

In one example of this concept, if you want to move an object from Earth's surface into space, you have to supply it with enough energy to climb out of Earth's gravitational well. One look at the fuel expended for any NASA rocket taking off and you are reminded of the energy needed to put the rocket's payload into deep space, far enough away that Earth is no longer trying to pull it back. So, to get there, the payload started from a state of negative gravitational potential energy and, after burning all that fuel, it ended in a state of zero gravitational potential energy relative to Earth itself.

In another example, imagine you are on a level field watching someone dig a hole and make a pile of dirt. If all you saw was the pile of dirt, the process would look miraculous—a small mound suddenly appeared out of nowhere. But once you see the hole

To reach zero gravity, a rocket burns fuel, going from negative potential gravitational energy toward zero, reached once it escapes Earth's pull. Here, an Atlas V lifts off, carrying NASA's Perseverance rover. Destination: Mars.

TRYON KEPT ON TRYING

Edward Tryon tells the story about how, during a lecture by a visiting physicist, he blurted out the question, "Could the universe be a vacuum fluctuation?" Everyone laughed, assuming he was making a joke, but people still read his paper.

from which it came, the miracle vanishes. The person with the shovel created a hole and a pile of dirt, but started with no hole and no pile of dirt. All this means is that you can create a universe that has zero total energy, but as long as you keep making holes and dirt piles in it, your universe can be immensely interesting.

Tryon closed his paper with one of our favorite quotes. Perhaps, he speculated, "our universe is simply one of those things that happen from time to time."

COSMOGENESIS

Let's work through a scenario for the origin of the universe—one most cosmologists embrace.

It starts with the quantum vacuum. The best way to visualize the pre–Big Bang universe is as a ball rolling down a hill. The higher the ball on the hill, the greater the potential energy it starts with. The very bottom of this hill is called the true vacuum state. If, on its descent, however, the ball nestled itself into a pothole on the hill, it now finds itself in what is called a false vacuum state. A lot of energy is stored in the false vacuum—one tiny nudge out of the hole and the ball continues on its own toward the true vacuum, releasing that potential energy. In the Newtonian world, the only way the ball can get out of the false vacuum is for someone to give it a push over the lip of the pothole. In the quantum world, however, there are several ways for it to escape without bothering to climb out. One of them is quantum tunneling, in which the thing in the pothole—a universe, for example—disappears and then instantly reappears outside of the hole to continue its slide down the hill.

Many competing ideas related to this scenario exist. All include a strong repulsive, antigravity pressure: something we see today as dark energy operates in the false vacuum. This is what powers the inflationary universe. Once the system arrives at the true vacuum, all the gravitational energy stored in the false vacuum must go somewhere, and in the inflationary scenario it triggers the fireball of particles and radiation we call the Big Bang.

You may have experienced this firsthand. Imagine a roller coaster where your cart starts high on a hill, chock-full of gravitational potential energy. While you descend, you gain speed as your potential energy converts to kinetic energy. This manifests as inflation in the pre–Big Bang universe. When you get to the bottom, there's usually extra track allowing you to overshoot and gracefully come to rest. But suppose instead there was just a brick wall waiting for you? On contact, all of your energy instantly converts to an explosion, killing everyone on board.

DEATH BY VACUUM DECAY

Every so often, cosmologists offer another hypothesis for how the universe might end. One of these hinges on the true and false vacuums that jump-started our universe. What if our current universe exists only in a false vacuum? A high-energy event could punch us out of the pothole, on a slide to the true bottom, unleashing a universe-ending fireball. In addition, our universe could theoretically tunnel through the side walls of the pothole, without provocation, and slide right on down to the true vacuum, also unleashing catastrophe as the true vacuum instantaneously obliterates us all—and everything else in the universe.

The good news is that the life span of a universe existing in a false bottom is expected to long outlive the current age of our universe—so you can rest easy tonight.

Calculations suggest that the energy in a false vacuum can be enormous—the energy contained in a cubic centimeter of false vacuum is greater than all the energy of the observable universe. That's plenty of energy to go around if you want to make more than one universe.

Oddly enough, in the early days of the inflation hypothesis, one of the difficulties was not trying to get inflation started, but rather to make it end. The challenge was politely dubbed the graceful exit problem. The energy diagram for this scenario resembles a simple hill, where the true vacuum at the bottom awaits you, and the slow roll down a shallow hill to get there is what allows a graceful exit after tunneling out of the pothole—the false vacuum.

BEFORE THE BIG BANG

We've now collected tools enough to grapple with another of those cosmically bedeviling questions: What existed before the Big Bang?

Some scientists consider attempts to answer or even pose a question like this to be complete nonsense. To quote the great non-astrophysicist St. Augustine, "The world was made not in time, but simultaneously with time. There is no time before the world." In other words, if time came into existence with the Big Bang, it makes no sense to talk about time before the Big Bang. It's like asking, "What is north of the North Pole?" No matter which direction you go from there, you're headed south. Even if you get into a helicopter, you will only rise above the pole, not north of it. It's not that there's nothing north of the North Pole; it's that there's not even nothing north of the North Pole. The premise of the question is inherently flawed.

When you think about it, in the universe of Pinocchio, with his proboscis lie detector, not all questions are valid either, with some leading to logical inconsistencies.

But the false vacuum universe scenario at least gives us a way of framing a disallowed question. Posed this way, the obvious answer to what was around before the Big Bang is "the quantum vacuum." This state presumably existed before the decay of the false vacuum.

But how would you verify such a claim? One way is to find something we can measure in the current universe that depended on the state of the universe before the Big Bang—which unfortunately amounts to a search for the smoke from a gun (and not even for the smoking gun itself). Unfortunately, inflation fundamentally prevents us from doing just that. How? Why?

The early universe underwent oscillations and expansion, loosely evoked in this abstract artwork.

Suppose we held in our hand a deflated balloon, wrinkled and twisted in all sorts of ways. If we inflated the balloon, a small enough creature, like an ant crawling on the surface, would think it's smooth and flat. Residents of Earth's surface know from space pictures and from other secondary methods that Earth is a sphere—but we are sufficiently small that locally, the land looks flat to our senses. (A good test of how flat Earth can look is to drive a car in any direction from Fargo, North Dakota.) After the balloon is inflated, the crinkles disappear, or are greatly diminished, regardless of how gnarly it started out. In the same way, when we look at the universe after inflation, any evidence from before has been stretched clean, so to speak, leaving us no way to determine how the universe began.

From the ground, Earth looks flat. From space—as in this view from
the International Space Station—the planet's curve is evident.

Even if we grant ourselves a vacuum energy beginning of the universe, another question lands on the table: Why did the false vacuum decay when it did? If the vacuum state existed for an infinite time prior, then why did it choose that moment 13.8 billion years ago to birth our universe instead of some other time? And that's as much of a philosophical and scientific frontier as anything else.

In the end, questions about what the universe was like before the universe began may be unanswerable, no matter whether it makes sense to ask the questions at all.

THE MULTIVERSE

With the widespread acceptance of the inflationary picture and the extraordinary success of quantum physics to help understand reality, however bizarre, we will assert that these two general truths must govern the origin of the universe. When you combine these two concepts, an extraordinary prediction emerges for free: the existence of other universes.

Let's return to the ball rolling down the hill. In a Newtonian world, the state of the ball is described by its position and velocity, both of which can be known at the same time to unlimited precision. In the quantum world, however, such a description is not possible because of the uncertainty principle. Consequently, the state of the ball must be described in terms of probabilities. Until the position of the ball is actually measured, we can think of it as existing in all the possible states, all at once.

So what does the ball descending from the false vacuum to the true vacuum look like in the quantum world? There is high probability that the ball will descend, but there's also a small probability that it hasn't left the pothole.

If you find this situation confusing, welcome to the club. The quantum world isn't like the familiar world we live in. And the universe is under no obligation to make sense to the human mind.

In an infinite ensemble of universes, all possibilities—no matter how improbable—will be realized somewhere. In this simple model of only two outcomes, there will be patches of the quantum vacuum where a Big Bang fireball was created, and other places where the system remains in a false vacuum. The result will be a collection of universes, each starting off at different times, each containing within it an observable universe like our own, and each separated from other universes by the rapidly inflating false vacuum. Inflation, in other words, is always going on somewhere: a phenomenon called eternal inflation.

This picture is called the multiverse. To envision it, imagine a large collection of bubbles that never touch each other because

THE BIG BOUNCE

Just as the false vacuum scenario provides a way to think about the question of what happened before the Big Bang, the Big Bounce is another pre–Big Bang hypothesis that allows us to contemplate the question. It proposes that after the universe reaches a certain point of expansion, it will eventually contract and condense into another infinitesimally small, hot mass that will trigger another rapid expansion. This hypothesis allows the universe to have been doing so from an infinitely long time ago, and to continue doing so for infinite time in the future: an endless cycle with no beginning and no end.

The multiverse—an infinite array of possible universes, ours only one of them—
may be like bubbles coexisting but never touching.

the inflation of the intervening space keeps them separated. In principle, the fundamental structure of any one universe—the very laws of physics themselves and the value of the fundamental constants of nature, such as the speed of light and the charge on an electron—could be different from that of all other universes. This fact provides us with possible answers to yet another vexing problem in cosmology: the so-called fine-tuning problem.

THE FINE-TUNING PROBLEM

What would our universe be like if the gravitational force were different? If it were stronger, gravity might pull everything back soon after the Big Bang, rendering its lifetime too short for stars, planets, and life to form. If gravity were weaker, on the other hand, matter might not collect into galaxies nor form stars or planets at all.

Or consider the charge on the electron. If it were a lot weaker than it is, atoms couldn't form. If it were a lot stronger, on the other hand, atoms might not exchange electrons to form molecules, so there would be no chemistry. In either case, life as we know it could not exist.

Both of these examples represent a fine-tuning of the universe. In fact, there's a cottage industry in the scientific community dedicated to figuring out how much wiggle room exists among the various constants of nature, yet still allowing life to develop. All these calculations point to a single conclusion: There is a very narrow range of allowed values for these constants if you want your universe to form life. Yet life has developed—how else could you be reading these words? Reconciling the narrow band of allowed values with the presence of life has come to be called the fine-tuning problem.

The multiverse saves the day.

If an infinite number of universes exist out there, each with different laws of physics and different constants of nature, then, just by chance, some of them will have the combination of laws and constants that fall within the narrow band that permits life. In those universes, life might also wonder why their universe is so finely tuned.

This solution to the fine-tuning problem is a cosmological version of what statisticians call the Texas sharpshooter argument. Imagine someone shooting bullets at random at the side of a barn. When the shooting is over, someone goes up, after the fact, and draws a bull's-eye circle around a cluster of bullet holes that happen to have landed near one another. You wouldn't do this and then compliment the shooter's accuracy. For the same reason, you shouldn't compliment our universe just because it happens to be among the very few that can form and sustain life. The bull's-eye searches for and positions itself on the only universe compatible with its existence.

A red-eared guenon monkey climbs the rainforest canopy of Equatorial Guinea. Finely tuned laws of physics enable the complex molecules necessary for the formation and evolution of life.

CLASSIFYING THE MULTIVERSE

Sooner or later, somebody was bound to devise a classification scheme to impose a little bit of order on this new frontier. In this case, that somebody was Swedish-American astrophysicist Max Tegmark. He suggests the following four categories of multiverses.

LEVEL 1 ❙ Between the edge of our observable universe and the larger, solo bubble universe encasing it are multiple other universes, not entirely unlike ours. We can't see them, and they can't see us. We exist outside of one another's realm. Level 1 multiverses are a little bit like scattered ships at sea, all sailing beyond each other's visible horizon. They can peer to their own horizons in every direction, but they are so far separated, they can't see one another at all. Meanwhile, they all coexist in the same ocean.

We know they come from the same physics stock as we do—our equations tell us so. The initial condition may be different, though. For example, some of them may contain different combinations of matter and energy. And some may have sufficient mass for their gravity to overcome their expansion, putting them in the Big Bounce club of universes. If there is an infinite number of these universes, then all combinations of matter, motion, and energy—all possibilities—are realized from one universe to the next. That means, for example, there is a universe out there with another version of you but reading a different book and sporting purple hair. In fact, there may be countless versions of you, some with your same memories, making all the decisions you wish you had made. The possibilities are endless. This is the multiverse familiar to science fiction readers: endless parallel universes.

And that's just the first of four levels.

LEVEL 2 ❙ A Level 2 multiverse contains many bubble universes. It's the kind of multiverse that eternal inflation generates. Here, bubble universes, each containing their own retinue of Level 1 universes, can express a different number of dimensions as well as different physical constants, completely changing the behavior and structure of matter and energy within them.

That being said, Level 2 universes are otherwise governed by the same laws and equations that we are. But now, with endless palettes of fundamental constants to choose from, we arrive at a simple solution to the fine-tuning problem: Just find the bubble universe whose physical constants allow life—and, within that, find the universe whose initial conditions gave rise to stars and planets. That may be the universe you live in.

LEVEL 3 | The Level 3 multiverse, often referred to as the many worlds hypothesis, is the side-by-side ensemble of all the Level 2 multiverses. In this construct, all quantum states are realized at all times as a branch point in time. In other words, every action and every decision within a world splits in that moment into other universes with different outcomes of that single action. In our universe, a single brain neuron firing might be the difference between one life-altering decision and another. So whatever memory you've conjured after reading that sentence, you can contemplate another "you" in a Level 3 parallel universe far, far away who made a different decision, whose neuron fired a different way, and whose life thereafter is completely different from yours.

LEVEL 4 | These are the multiverses associated with all possible mathematical structures. In a Level 4 universe, Newton's laws might take on many different forms. For example, gravity might not depend on the mass of the objects exerting a gravitational force. Or there are places where systems naturally become more ordered over time. That would be freaky. Observing events in that universe would be like watching a film run backward. Omelets would naturally deconstruct and revert into eggs and piles of cheese. Shards of a shattered cup on the floor would spontaneously reassemble and jump back up on the table, poised to be filled with hot chocolate.

We tried hard, but the full bag of tricks contained within a Level 4 multiverse lies far beyond our physical or even philosophical abilities to visualize or even analogize.

IS THIS REALLY SCIENCE?

Sometimes theoretical physics can become so difficult to comprehend with our humble human brains and minuscule life spans relative to the universe that philosophers step in and give it a go. You cannot read very far into the multiverse literature before you start seeing references to 14th-century philosopher and theologian William of Ockham, who is credited with declaring that "Multiplicity ought not be posited without necessity." If you're into Latin, he said it this way: *"Pluralitas non est ponenda sine necessitate."* And if he were around today, he surely would have invented the KISS principle: "Keep it simple, stupid!" Now known as Ockham's razor, his point is a thoroughly embraced reality check on the complexity of our ideas.

Have we gone too far? Too much conjecture? Too much philosophizing? All four levels of multiverse we've talked about involve systems with which we cannot, even in principle, communicate. If the central tenet of the scientific method is that all propositions must be tested against experiment and observation, then how can statements about the multiverse be part of science?

And let's face it—the farther away we get from the Level 1 multiverse, the more reality fades into the shadows of our ignorance.

Critics of the multiverse idea point out that the predictions we can make about other universes using our best particle theories can never be tested—again concluding that it is not real science. Supporters, on the other hand, point out that the theory is very well supported, founded on quantum mechanics and the inflationary universe. Furthermore, it is not necessary that every prediction of a theory be verified before the theory is accepted. In the 1920s, for example, the physics community accepted general relativity based on only two tests: oddities in the orbit of Mercury and the bending of starlight during the 1919 eclipse of the Sun. All the other triumphs of the theory—the existence

With select astronomical events—like this solar eclipse, shown as a time-lapse— scientists can test their ideas. A 1919 eclipse provided the daytime darkness to support Einstein's new theory for how the Sun's gravity bends light.

FOR THE BIRDS

Physicist Richard Feynman is known to have said that "The philosophy of science is about as much use to scientists as ornithology is to birds."

of black holes, gravitational red shift, and gravitational waves, for example—came decades later.

As it happens, theoretical physicists—always a clever lot—have already started thinking of ways to detect phenomena in our own observable universe that would reveal the existence of other universes. To take one example, if our own universe happened to suffer a glancing collision with another universe in our distant past, characteristic imprints of that encounter may reveal themselves in the map of the cosmic microwave background. They're still looking.

Many questions about our own humble universe—perhaps just one of an infinite number of universes—remain to be answered. We implore you, dear reader, to continue to stoke your curiosity and to ask the impossible cosmic questions. For the goal of our short life is not to find the answers, but to search for new places to stand so that we can formulate questions not previously imagined. Along that journey, as you shape your own cosmic perspective, we bid you, as always, to keep looking up.

Amid these myriad observations, calculations, technologies, hypotheses, and theories, let us take a moment to look up in perfect silence at the stars.

ACKNOWLEDGMENTS

What a delight to work with the editors and designers of National Geographic Books.

A secret fear of authors is a publisher's urge to change the vision of what you've so carefully conceived. But the folks at Nat Geo really know what they're doing. They know how to help you say what you mean and mean what you say. But, more important, their creativity elevates words on the page with design and image elements that enable the final book to shine in ways undreamed of by the authors. That's collaboration at its finest. That's *StarTalk* elevated by National Geographic. Specifically, executive editor Hilary Black, senior editor Susan Hitchcock, and associate editor Moriah Petty were there at every turn, supporting and guiding our writing efforts, and being responsive to the *StarTalk* brand where and when the content called for it, while copy editor Heather McElwain, senior production editor Judith Klein, and managing editor Jennifer Thornton ensured that the book's conversational style didn't float too far away from literary norms. Not only that, creative director Melissa Farris, art director Sanaa Akkach, director of photography Susan Blair, and photo editor Adrian Coakley continued the Nat Geo legacy of enhancing the readers' experience through compelling visuals that, in this case, help to bring the universe down to Earth.

Previous pages: The dazzling heart of our Milky Way galaxy

FURTHER READING

CHAPTER 1

Koestler, Arthur. *The Sleepwalkers: A History of Man's Changing Vision of the Universe.* Penguin, 1990.

Sobel, Dava. *Glass Universe: How the Ladies of the Harvard Observatory Took the Measure of the Stars.* Viking, 2016.

Tyson, Neil deGrasse. "Stick-in-the-Mud Science." *Natural History* 112, no. 2 (2003): 32+.

Webb, Stephen. *Measuring the Universe: The Cosmological Distance Ladder.* Springer-Praxis, 1999.

CHAPTER 2

Hawkins, Gerald, and John B. White. *Stonehenge Decoded.* Hippocrene Books, 1988.

Levin, Janna. *Black Hole Blues and Other Songs from Outer Space.* Knopf, 2016.

Magli, Giulio. *Archaeoastronomy: Introduction to the Science of Stars and Stones.* Springer, 2016.

Selin, Helaine, ed. *Astronomy Across Cultures: The History of Non-Western Astronomy.* Springer, 2000.

CHAPTER 3

Randall, Lisa. *Dark Matter and the Dinosaurs: The Astounding Interconnected-ness of the Universe.* HarperCollins, 2015.

Rubin, Vera. *Bright Galaxies, Dark Matters.* Springer-Verlag, 1996.

Stern, Alan, and David Grinspoon. *Chasing New Horizons: Inside the Epic First Mission to Pluto.* Picador, 2018.

Stern, S. A., et al. "Overview of Initial Results from the Reconnaissance Flyby of a Kuiper Belt Planetesimal: 2014 MU_{69}." Available online at arxiv .org/pdf/1901.02578.pdf.

Tyson, Neil deGrasse, and Donald Goldsmith. *Origins: Fourteen Billion Years of Cosmic Evolution.* W. W. Norton, 2004.

Williams, Jonathan P., and Lucas A. Cieza. "Protoplanetary Disks and their Evolution." *Annual Review of Astronomy and Astrophysics* 49, no. 1 (2011): 67–117.

CHAPTER 4

Balbi, Amedeo. *The Music of the Big Bang: The Cosmic Microwave Background and the New Cosmology.* Springer, 2008.

Guth, Alan. *The Inflationary Universe: The Quest for a New Theory of Cosmic Origins.* Basic Books, 1988.

Riess, Adam G., et al. "Observational Evidence from Supernovae for an Accelerating Universe and a Cosmological Constant." Available online at iopscience.iop.org/article/10.1086/300499/pdf.

CHAPTER 5

Bartusiak, Marcia. *Einstein's Unfinished Symphony: Listening to the Sounds of Space-Time.* Joseph Henry Press, 2000.

Feynman, Richard P., and Steven Weinberg. *Elementary Particles and the Laws of Physics: The 1986 Dirac Memorial Lectures.* Cambridge University Press, 1987.

Greene, Brian. *The Elegant Universe: Superstrings, Hidden Dimensions, and the Quest for the Ultimate Theory.* W. W. Norton, 2003.

Riordan, Michael. *The Hunting of the Quark: A True Story of Modern Physics.* Simon & Schuster, 1987.

Tegmark, Max, and John Archibald Wheeler. "100 Years of Quantum Mysteries." Available online at space.mit.edu/home/tegmark/PDF/quantum.pdf.

CHAPTER 6

Bostrom, Nick. "Ethical Issues in Advanced Artificial Intelligence." Available online at www.fhi.ox.ac.uk/wp-content/uploads/ethical-issues-in -advanced-ai.pdf.

Dodd, Matthew S., et al. "Evidence for Early Life in Earth's Oldest Hydrothermal Vent Precipitates." *Nature* 543 (2017): 60–64.

Koshland, Daniel E., Jr. "The Seven Pillars of Life." *Science* 295, no. 5563 (2002): 2215–16. Available online at science.sciencemag.org/ content/295/5563/2215/tab-pdf.

Kurzweil, Ray. *The Singularity Is Near: When Humans Transcend Biology.* Viking, 2005.

CHAPTER 7

Hand, Kevin Peter. *Alien Oceans: The Search for Life in the Depths of Space.* Princeton University Press, 2020.

McKay, Chris P. "What Is Life—and How Do We Search for It in Other Worlds?" *PLoS Biology* 2, no. 9 (2004): 260–63. Available online at www.ncbi .nlm.nih.gov/pmc/articles/PMC516796/pdf/pbio.0020302.pdf.

Scoles, Sarah. *Making Contact: Jill Tarter and the Search for Extraterrestrial Intelligence.* Pegasus Books, 2000.

Trefil, James, and Michael Summers. *Imagined Life: A Speculative Scientific Journey among the Exoplanets in Search of Intelligent Aliens, Ice Creatures, and Supergravity Animals.* Smithsonian Books, 2019.

CHAPTER 8

Borissov, Guennadi. *The Story of Antimatter: Matter's Vanished Twin.* World Scientific Publishing, 2018.

Feynman, Richard. *QED: The Strange Theory of Light and Matter.* Princeton University Press, 1986.

Greenstein, George, and Arthur Zajonc. *The Quantum Challenge: Modern Research on the Foundations of Quantum Mechanics.* Jones & Bartlett Learning, 2005.

CHAPTER 9

Levin, Janna, Evan Scannapieco, and Joseph Silk. "The Topology of the Universe: The Biggest Manifold of Them All." *Classical and Quantum Gravity* 15 (1998): 2689–98.

Oppenheimer, Clive. "Climatic, Environmental And Human Consequences of the Largest Known Historic Eruption: Tambora Volcano (Indonesia) 1815." *Progress in Physical Geography: Earth and Environment* 27, no. 2 (2003): 230–59.

Schmidt, Nikola, ed. *Planetary Defense: Global Collaboration for Defending Earth from Asteroids and Comets.* Springer, 2019.

CHAPTER 10

Bojowald, Martin. "What Happened Before the Big Bang?" *Nature Physics* 3 (2007): 523–25. Available online at www.nature.com/articles/nphys654.pdf.

Tegmark, Max. "Parallel Universes." *Scientific American* (March 2003): 40–51. Available online at space.mit.edu/home/tegmark/PDF/multiverse_sciam.pdf.

Tegmark, Max, and Nick Bostrom. "Is a Doomsday Catastrophe Likely?" *Nature* 438 (2005): 754. Available online at www.nature.com/articles/438754a.

ILLUSTRATIONS CREDITS

JPL-Caltech; 64, Christian Offenberg/Alamy Stock Photo; 65, NSF/ LIGO/Sonoma State University/A. Simonnet; 67, Dave Yoder/National Geographic Image Collection; 68, NASA/MSFC/David Higginbotham/ Emmett Given; 71, ESO/L. Calçada; 72-3, Moonrunner Design/National Geographic Image Collection; 74, SPL/Science Source; 79, Henning Dalhoff/Bonnier Publications/Science Source; 80, William Turner/ Getty Images; 83, Courtesy Carnegie Institution for Science Department of Terrestrial Magnetism Archives; 87, NASA, ESA and M. Livio and the Hubble 20th Anniversary Team (STScI); 89, NASA, ESA, J. Debes (STScI), H. Jang-Condell (University of Wyoming), A. Weinberger (Carnegie Institution of Washington), A. Roberge (Goddard Space Flight Center), G. Schneider (University of Arizona/Steward Observatory), and A. Feild (STScI/AURA); 92, Lynette Cook/Science Source; 94, NASA/ Johns Hopkins University Applied Physics Laboratory/Southwest Research Institute; 96, Detlev van Ravenswaay/Science Source; 98-9, Adolf Schaller for STScI; 100, Courtesy KIPAC. Simulation: John Wise, Tom Abel; Visualization: Ralf Kaehler; 104, NASA; 106, ESA and the Planck Collaboration; 109, David Parker/Science Source; 112, ESA-D. Ducros, 2013; 113, ESA/Gaia/DPAC; 114, David A. Hardy/Science Source; 116, NASA's Goddard Space Flight Center; 119, Maximilien Brice, CERN/ Science Source; 124, Ken Eward; 126-7, Pasieka/Science Source; 128, NASA, ESA and H. Bond (STScI); 131, aluxum/Getty Images; 137, David Parker/Science Source; 138, Jose Antonio Penas/Science Source; 142, Science & Society Picture Library/Getty Images; 143, NYPL/Science Source; 144, David Parker/Science Source; 148, Science & Society Picture Library/Getty Images; 149, Courtesy of *Particle Fever*; 152-3, IKELOS GmbH/Dr. Christopher B. Jackson/Science Source; 154, The Picture Art Collection/Alamy Stock Photo; 157, Roger Ressmeyer/Corbis/VCG via Getty Images; 159, Lynette Cook/Science Source; 161, NASA Photo/ Alamy Stock Photo; 162, Keith Chambers/Science Source; 165, Steve Gschmeissner/Science Source; 168, Greg Lecoeur/National Geographic Image Collection; 171, Philippe Psaila/Science Source; 175, Mark Garlick/Science Source; 177, NOAA Okeanos Explorer Program/Science Source; 178, Eye of Science/Science Source; 180-81, Babak Tafreshi/ National Geographic Image Collection; 182, NASA/JPL-Caltech; 185, NASA/JPL-Caltech; 187, Moviestore Collection Ltd/Alamy Stock Photo; 188, Lowell Observatory Archives; 189, ESA/DLR/FU Berlin; 190, Dr. Seth Shostak/Science Source; 192, Zoediak/Getty Images; 193, NASA/

JPL-Caltech/MSSS; 194, Chris Butler/Science Source; 196, Frans Lanting/MINT Images/Science Source; 199, Courtesy of Lucasfilm Ltd. STAR WARS© & ™ Lucasfilm Ltd.; 201, Courtesy of the Ohio History Connection, #AL07146; 204, NASA/JPL-Caltech; 206, Bettmann/Getty Images; 209, Mark Garlick/Science Source; 211, Lynette Cook/Science Source; 212-13, agsandrew/Shutterstock; 214, Illustris Collaboration via ESO; 216, Henning Dalhoff/Bonnier Publications/Science Source; 218, Keystone-France\Gamma-Rapho via Getty Images; 223, Richard Kail/Science Source; 226, CERN, Maximilien Brice and Julien Marius Ordan/Science Source; 228, AIP Emilio Segrè Visual Archives, Margrethe Bohr Collection/Science Source; 231, Carol and Mike Werner/Science Source; 234, NASA, ESA, N. Smith (University of California, Berkeley), and The Hubble Heritage Team (STScI/AURA); 236-7, Mark Stevenson/Stocktrek Images/Science Source; 238, Detlev van Ravenswaay/Science Source; 242, Charles Preppernau; C. R. Scotese PALEOMAP Project; 245, Neil deGrasse Tyson; 246, Michael Nichols/National Geographic Image Collection; 248, Alan Copson/Jon Arnold Images Ltd/Alamy Stock Photo; 252, Spencer Sutton/Science Source; 255, Robert Clark/National Geographic Image Collection; 258, Tomasz Dabrowski/Stocktrek Images/National Geographic Image Collection; 261, Mikkel Juul Jensen/Science Source; 262, Paul Zizka/Cavan Images; 264-5, Dr. Dieter Willasch (astro-cabinet.com); 266, Fermilab; 269, Private Collection/Bridgeman Images; 271, James King-Holmes/Science Source; 273, AP Photo/John Raoux; 277, Jen Stark/"Abyss" (detail), 2011, acid-free hand-cut paper, wood, acid-free foamcore, glue, light, 20 x 20 x 33 in; 278, NASA photo edited by Stuart Rankin; 281, Detlev van Ravenswaay/Science Source; 283, Tim Laman/NPL/Minden Pictures; 287, Philip Hart/Stocktrek Images/National Geographic Image Collection; 288, Paranyu Pithayarungsarit/Getty Images; 290-1, NASA, ESA, and T. Brown (STScI), W. Clarkson (University of Michigan-Dearborn), and A. Calamida and K. Sahu (STScI).

INDEX

Boldface indicates illustrations.

ABOUT THE AUTHORS

NEIL DEGRASSE TYSON is an astrophysicist and Frederick P. Rose Director of the Hayden Planetarium at New York's American Museum of Natural History. He is the author of more than a dozen books, many of them international bestsellers, and numerous articles, both scholarly and for the general public. He is the host of *StarTalk,* a podcast, and of two seasons of *Cosmos,* televised by Fox and National Geographic. He has received 21 honorary doctorates as well as NASA's Distinguished Public Service Medal. He and his wife live in New York City.

JAMES TREFIL, Clarence J. Robinson Professor of Physics at George Mason University, is internationally recognized as an expert in making complex scientific ideas comprehensible. He is the author of numerous magazine articles and books on science for the general public, including National Geographic's *Space Atlas* and *The Story of Innovation.* He and his wife live in Fairfax, Virginia.

I EXPLORE OUR UNIVERSE

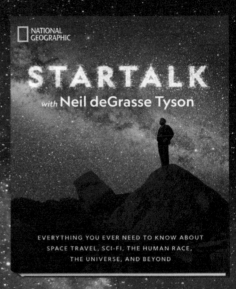

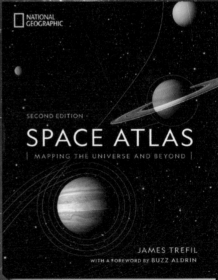